AERO-ENGINED RACING CARS AT BROOKLANDS

By the same author:

The 200 Mile Race (1947)

The Story of Brooklands, Vol. 1 (1948)

The Story of Brooklands, Vol. 2 (1949)

The Story of Brooklands, Vol. 3 (1950)

The World's Land Speed Record (1951, 1964)

Continental Sports Cars (1951, 1952)

The History of Brooklands Motor Course (1957, 1979)

The Bugatti Story (1960) USA

Montlhéry—The History of the Paris Autodrome (1961)

The Sports Car Pocketbook (1961)

The Vintage Years of the Morgan Three-Wheeler (1970)

Motor Sport Book of the Austin Seven (Editor) (1972)

Motor Sport Book of Donington (Editor) (1973)

The History of Motor Racing (1977)

My 30 Years of Motoring for *Motor Sport* (1982)

Volkswagen Beetle (1982)

Mercedes-Benz 300SL (1983)

Vintage Motor Cars (1985)

Brooklands Grants (1995)

AERO-ENGINED RACING CARS AT BROOKLANDS

BILL BODDY

First published in 1992
Reprinted 1996

British Library Cataloguing-in-Publication Data:
A catalogue record for this book is available from the British Library.

ISBN 0 85429 867 3

Library of Congress catalog card no. 92-85353

G. T. Foulis & Company is an imprint of
Haynes Publishing, Sparkford,
Nr. Yeovil, Somerset, BA22 7JJ, England.

Typeset by Wyvern Typesetting Ltd., Bristol
Printed in Great Britain by Hillman Printers (Frome) Ltd,
Handlemaker Road, Marston Trading Estate, Frome, Somerset BA11 4RW.

Contents

Acknowledgements

It is impossible, and it could be invidious to try, to thank all those who over so many years have helped me acquire information about the cars this book covers.

The number of people who so kindly replied to my letters, sent me data and photographs, let me study their old motoring-orientated albums and talked to me about their cars, is so large that individual mention defeats me. From such help I was assisted materially in writing my *History of Brooklands Motor Course* (Grenville Publishing Company 1957/1979) of which this more specialized work is something of a spin-off.

However, I would like to thank Lord Montagu of Beaulieu for allowing me to use material I wrote for his late-lamented *Veteran & Vintage Magazine*, those like Peter Wike, Clive Windsor-Richards, Graeme Simpson, and others from whose letters I have quoted, David Wilson, whose talented research into the Zborowski family has filled in the odd item, and Terence Cuneo for allowing his splendid painting of the first of the Brooklands' aero-engined racing cars (the 9-litre V12 Sunbeam, seen coming at speed off the Members' banking into the Railway Straight) to feature on the dust-jacket.

Along the years so many people have very kindly sent me motor-racing photographs, some of which I have used in this book, pictures which have passed into the 'Boddy Collection' and which, again, it is impossible to credit individually. If this causes any ill-feeling I offer my sincere apologies, and my sincere thanks for the interest shown.

Finally, I would like to thank Mrs Elaine Peberdy for making such an excellent job of typing up the manuscript from my rough and ready copy.

Bill Boddy
Radnorshire, 1992

Introduction

Some explanation is necessary for a book on such a specialized subject as aero-engined racing cars, but I make no excuse for writing it!

From the time when I began to read about Brooklands and then made my first visit there to a race meeting—with my war-widowed mother, who perhaps wanted to see if this place where bookmakers operated was suitable for a boy of 14—it was the monster racing cars, that used the banked outer circuit as their stamping ground, that particularly fascinated me. Not many of them were very scientific. No matter; they were fast, individualistic, exciting! In this book I have attempted to trace in some detail their separate careers.

Why not include all the Brooklands giants, you may ask? Cars like the Blitzen Benz and the 200 hp Benz four-seater, the 15-litre 1912 Lorraine-Dietrich *Vieux Charles Trois*, the various big Mercedes that made Brooklands their home, likewise the gigantic Fiats, the old 10.5-litre Delage and others of that kind. Apart from having to set a limit somewhere, none of these exciting cars was built expressly with Brooklands racing in mind. The big Benz cars came to England after hill-climb and Land Speed Record bids, the Lorraine 'Old Charles the Third' was built for the 1912 French Grand Prix, the Mercedes family were originally mostly road-racing cars, as were the Fiats, and the Delage started as a hill-climb competitor and Land Speed Record car.

Aeroplane engines had their principal usage on the ground in Land Speed Record contenders, apart from the Brooklands cars so endowed, and those that ran demonstrations at the Weybridge track are mentioned in a separate chapter. With a few exceptions, those aero-engined Brooklands racing cars were constructed by amateurs, who saw that this was an inexpensive way of obtaining real power, and that an old chassis so powered could be as fast as expensive factory-built cars. Such engines were available after the First World War from Government disposal stores for the proverbial song.

For example, the Aircraft Disposals Board was offering new 160 hp Mercedes engines, presumably acquired as war reparations, for £60, or for £30 in reconditioned form, and a Wolseley Viper vee-eight, which had cost £1,500 when new, for £75. There was no shortage of such bargain-price power units. At Waddon (later Croydon) Aerodrome in Surrey no fewer than 30,000 RAF engines were being advertised for sale in 1920. Another source of supply was Coley's of Kingston, where it was said that old Fiat aviation engines cost £25 each, and spark plugs for such engines 2/6d (12½p) a dozen. A story, probably apocryphal, was circulating to the effect that even the low prices asked for these surplus engines could be reduced further by the judicious use of the tang-end of a file on a vulnerable part of the machinery, such as a water-jacket, when no-one was looking, the engine then being marked down as damaged when attention was drawn to the defect!

Thus it was tempting to put such inexpensive power units, of anything from 150 to over 300 b.h.p., into an old chassis with a suitably long wheelbase, for handicap racing at the Weybridge Track. Ken Kirton, the Honiton old-car fancier, had a theory that such monsters were sometimes hampered by having gear ratios that were too low. An aged chain-drive chassis had the advantage here, because its final-drive ratio could be easily experimented with by altering the size of the chain sprockets.

If anyone regrets the exclusion from this book of the aforesaid non-aero-engined racing giants, I, nevertheless, think I have made the right decision. It was the aero-engined cars, such as the famous 350 hp V12 Sunbeam and Count Zborowski's legendary 'Chitty-Bang-Bangs' that truly captured the imagination of the post-Armistice Brooklands spectators. In the end, it was the aero-engined Napier-Railton which gained for all-time the Brooklands lap-record. And when the top lap-speeds in the vintage years (pre-1931) for the conventional versus the aeroplane-engined monsters are compared, the latter are seen to be the equal of most conventional giants, as the following tables show:

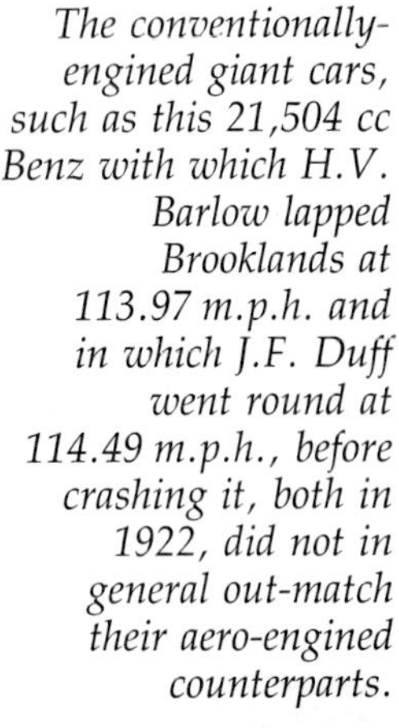

The conventionally-engined giant cars, such as this 21,504 cc Benz with which H.V. Barlow lapped Brooklands at 113.97 m.p.h. and in which J.F. Duff went round at 114.49 m.p.h., before crashing it, both in 1922, did not in general out-match their aero-engined counterparts.

Non-aero-engined Class-A (over 8,000 cc) cars

	m.p.h.
21.5-litre Benz (C. Paul)	115.55
10.5-litre Delage (J. Cobb)	132.11
21-litre Fiat (F. Nazzaro)	121.64
10-litre Fiat (J. Cobb)	113.45
15-litre Lorraine-Dietrich (W. Hawkes)	105.97
12.8-litre Mercedes (C. Lane)	105.02
Average speed	115.62

Aero-engined cars

	m.p.h.
18.3-litre Sunbeam (K. Lee Guinness)	122.67
23-litre 'Chitty-Bang-Bang' (Count Zborowski)	113.45
18.8-litre 'Chitty-Bang-Bang' (Count Zborowski)	108.27
21.7-litre Fiat (E. Eldridge)	123.89
20.5-litre Isotta-Maybach (L. Champion)	114.75
14.7-litre Mercedes (Count Zborowski)	112.68
27-litre Thomas Special (Parry Thomas)	125.77
11.7-litre Wolseley-Viper (K. Don)	112.68
Average speed	116.77

These tables actually favour the conventionally-powered cars, because I have allowed Felice Nazzaro his electrically-timed lap at over 121 m.p.h. in 1908, whereas many people consider A.V. Ebblewhite's hand-timed lap at 107.98 m.p.h. to be more credible, and I have not allowed Ernest Eldridge to lap at 125.45 m.p.h. in the Fiat 'Mephistopheles' which was established in a club match race but was not officially recognized.

Whichever way you look at it, these old aero-engined cars were fascinating, fast, and even dangerous, as you will see. Unfortunately, two people were killed during practice laps involving them. The big Sunbeam twice took Harry Hawker off the track, on the second occasion through the corrugated-iron Railway Straight fence when a tyre burst, and the same cause sent Count Zborowski off the same part of the Track and into the ditch in 'Chitty-I'. The Wolseley Viper nearly took Kaye Don and his passenger over the top of the Members' banking when its brakes proved inadequate at the end of a race. John Cobb said, after quick laps in the Napier Railton, that it was like seeing how far one could lean out of a window without actually falling out.

In the days when these great cars were in the news it was a wonderful time in which to go down to the Surrey Motor Course to see them in action. It seems a long time ago, a different age; one that will never return. Brooklands as a race track was destroyed during the last World War, and in any case the Brooklands authorities became fearful of the older racing cars after 1930—metal fatigue and all that—although some of the cars this book deals with persisted after that; the Napier Railton right up to 1937. I hope you will derive some enjoyment from my recalling those now faraway days.

EXPLANATION

To understand properly what follows it is necessary to know something about Brooklands Track, which H.F. Locke King built at his own expense—some £150,000 in the currency of the time—on his estate at Weybridge in 1906/07. This concrete-surfaced Motor Course, the very first of its kind in the world, was far from being a 'speed bowl' as sometimes described. Its two curves, the Members' and the Byfleet, were banked to allow drivers to lap, theoretically at 120 m.p.h. with their hands off the steering-wheel. To make this possible these bankings were so steep that it was almost impossible for a person to climb up them.

The lower of these banked curves was at the Byfleet side of the course, and this joined the shorter, steeper Members' banking by means of an unbanked reverse-curve at the Fork, where the 991–yards long Finishing Straight ran up to join the main circuit at the foot of the aforesaid Members' banking. This reverse-curve was quite difficult for the faster cars to negotiate. The Track had, in fact, been built on horse-racing lines. The one-acre Paddock with covered bays for the competitors' cars, a fine Clubhouse, and the Telegraph-board, was on the left, going up the straight.

The two banked curves were joined at the eastern side by the half-mile Railway Straight, which ran below the embankment of the L. & S. W. R. The lap distance round the outer-circuit was 2 m 1,350 yds. It was over this fast circuit that most of the early Brooklands Automobile Club races were run. In order that all kinds of cars, from stripped sports cars and specials to the latest racing cars could compete together, these races were invariably run on handicap, to starting times worked out by the indefatigable Mr A.V. Ebblewhite, based on known performance and practice lap speeds. Most of the aero-engined cars would be heavily handicapped, starting on or close to the scratch mark.

These handicap races were normally divided into Short and Long Handicap contests, the former of around 5.75 miles, the latter of about 8.5 miles, with the entries grouped roughly in speed-categories, and the

Brooklands Track, where all the action took place between 1907 and 1939. The Byfleet banking and aeroplane sheds are in the foreground; the Fork, Finishing Straight and the Members' banking are on the far side.

handicapping sorting out the finer divisions. Thus there were 75 m.p.h., 90 m.p.h. and 100 m.p.h. Short and Long Handicap races, with the very fastest, most exciting cars set to run in the delightfully-named 'Lightning Handicaps'. The speeds quoted in the following chapters are those issued by the official B.A.R.C. timekeepers.* Further information will be found in my *History of Brooklands Motor Course* (Grenville, London, 1957/1979).

*Brooklands lap-speeds and record-bid average speeds have been compiled from the official records of the Brooklands Automobile Racing Club, endorsed by timekeepers A.V. Ebblewhite and T.D. Duttom, etc., which I am fortunate to have in my possession. If they differ from those in other books this may be because some authors have had to rely on contemporary Press accounts.

1. The 1913 V12 Sunbeam

The first use of an aeroplane engine in a Brooklands racing car goes to the credit of Louis Coatalen, the famous Chief Engineer of the Sunbeam Motor Car Company of Wolverhampton, Staffordshire. The Sunbeam Company had been successful in the pre-1914 period of racing and record breaking, one of the cars used for this purpose being their standard 25/30 hp chassis fitted with a single-seater body and known as 'Toodles IV'. Using the same type of chassis but seeking more speed, Coatalen then installed in one of these cars a side-valve vee-twelve cylinder aeroplane engine of 80×150 mm. (9,048 cc) which was destined to be developed into the Sunbeam Mohawk aero-engine which saw useful service in the coming war. ('Toodles IV' was the six-cylinder Sunbeam. The V12 was known as 'Toodles V'.)

By this method the Sunbeam Company not only gained good publicity on the Track, but Coatalen was able to develop the engine far more effectively than if it had been confined to experimental flying, because not only were aeroplanes not very practical in 1913 but weather conditions could seriously delay a test programme. So, into a 25/30 hp chassis this powerful engine went. It gave Sunbeam the fastest car on Brooklands at that time, the power developed being around 200 b.h.p. at 2,400 r.p.m. in a car of no great weight; indeed, the rear sections of the side-members were lead-filled, to aid wheel adhesion.

There is no doubt that Coatalen intended this 9-litre engine for aviation work. Non-adjustable tappets were used, permissible as a weight-saving item on an aero-engine which would be subjected to frequent inspection and rebuilding when required. The two banks of cylinders were at an inclined angle of 60 degrees, the camshaft for the side-by-side valves situated between them. The cylinders were in blocks of three and the two exhaust manifolds occupied most of the space between the two banks of cylinders. Carburation for such engines presented a problem, and at first very long induction pipes were tolerated, fed by two carburettors beneath the car's scuttle. This presumably proved unsatisfactory, and at a later stage each cylinder block was given a vertical Claudel Hobson carburettor, each on its own inlet manifold.

The big-ends were mounted side-by-side on the crankpins, permitted by slightly staggering the cylinders on the crankcase. An innovation was dry-sump lubrication, using a pressure and a scavenge pump. The oil tank was in the car's tail, oil being led along big diameter pipes outside the chassis frame, which cooled the lubricant when the Sunbeam was running. A further inducement to cool oil was a ribbed copper base chamber below the crankshaft. Ignition was by a magneto for each side of

First of the aero-engined giants—the 9-litre V12 single-seater raced by the Sunbeam Motor Car Co. of Wolverhampton during 1913/14. It is seen here on the Members' banking.

the engine, a cross-shaft driven by skew gears from the front of the camshaft providing the actuation.

Louis Coatalen must have been eager to see how such a powerful engine would serve in a racing car. It was installed in a sub-frame mounted at three points on the chassis. The drive went *via* a cone clutch and open propeller-shaft, with torque taken by a pressed steel torque-arm, to a back axle having the requisite high ratio of 2.0 to 1 and no differential. (Coatalen did not use differentials on his Grand Prix racing cars but did so on his production models, apart from on the little 8/18 hp Talbot and 8 hp Talbot-Darracq, which caused some controversy in the motor papers in post-war times.)

The 9-litre Sunbeam ran on 880×120 Palmer cord tyres on steel artillery wheels, and its wheelbase measured 10′6″. It was to make its first public appearance at the 1913 B.A.R.C. August meeting, the celebrated French-ace, Jean Chassagne, being entrusted to drive it. Alas, the clutch had given trouble and it was posted a non-runner. The Brooklands spectators ('The Right Crowd and No Crowding', as the Track's later slogan boasted) had to wait until the autumn meeting to see what the Sunbeam could do. Chassagne drove it and was third in the 100 m.p.h. Short Handicap race, a cautious start (that clutch?) and difficulty in deciding where, if at all, he could overtake, placing him ahead of Percy Lambert's 4.7-litre Talbot which had started from scratch with him, but vanquished by the winning Coupe de *L'Auto* Sunbeam of C.A. Bird, and N. Hind's 6.5-litre Berliet 'Black Beetle'. However, it was an auspicious debut for the aero-engined Sunbeam, for Chassagne had done his fastest lap at 114.49 m.p.h.

The Sunbeam and the Talbot met again in the Autumn 100 m.p.h. Long Handicap race. This time Chassagne passed below Lambert coming off the Members' banking and swept on to win the fastest Brooklands race up to that time, at 110 m.p.h., the Sunbeam's flying laps being at 115.82 and 118.58 m.p.h. Clearly it was a very fast car, the streamlining aided by faired-in front dumb irons which came to a point, and a very slim cockpit and slender tail. Incidentally, the French driver had to wear jockey's 'silks', consisting of grey coat and sleeves and a blue cap with white stars, for purposes of identification, racing numbers not having been introduced at Brooklands at that time.

The pre-war aero-engined Sunbeam had proved very effective, having, in fact, made the fastest Brooklands' lap, if one discounts the controversial electrically-timed speed of 121.64 m.p.h. by Nazzaro's Fiat during the 1908 Match Race against Newton on the Napier. But Coatalen had a bigger fish to fry. The World's one-hour record was a coveted target. In 1913 little Percy Lambert had set it to 103.84 m.p.h. with the side-valve 25 hp Talbot. Tyres were a limiting factor at that time in breaking the record, but nevertheless Jules Goux came to Brooklands and lifted it to 106.2 m.p.h. with one of the 1912 5.6-litre Grand Prix Peugeots. The Sunbeam obviously had an excellent chance of beating this speed for the hour run, if its tyres would last the distance, and it proved sufficiently reliable.

In October 1913 Chassagne was ready, but although he kept the Sunbeam's pace down to 112/115 m.p.h., it was too much, and after about 38 miles the tread of the tyre on the offside rear wheel was flung off. The attempt was abandoned. Two days later, on 11 October, with security bolts for the Palmer Cord tyres, the car was set up for another attempt on the hour record. As the scavenge oil pump was not very effective, excess oil was drained from the base chamber before Chassagne set off. After a fast standing-start lap he eased back to regular circuits at just above 107 m.p.h., and this time a new hour record of 107.95 m.p.h. was established.

In fact, Coatalen allowed the Sunbeam to continue, with a rapid change of tyres at his depot at the top of the Finishing Straight, and further World and Brooklands records were secured, as follows:

World 50 mile record	108.39 m.p.h.
World 100 mile record	107.93 m.p.h.
World 150 mile record	105.57 m.p.h.
World 50 kilo record	174.90 k.p.h. (109.30 m.p.h.)
World 100 kilo record	174.29 k.p.h. (108.93 m.p.h.)
World 150 kilo record	173.95 k.p.h. (108.72 m.p.h.)
World 200 kilo record	169.85 k.p.h. (106.16 m.p.h.)
World 250 kilo record	156.71 k.p.h. (97.94 m.p.h.)
World one hour record	107.95 m.p.h.
Brooklands 60 RAC Class f.s. 0.5 mile record	118.66 m.p.h.
Brooklands 60 RAC Class f.s. mile record	120.73 m.p.h.
Brooklands 60 RAC Class s.s. 10-laps record	110.03 m.p.h.

(The Sunbeam was rated by the RAC formula at 47.6 hp. It also established 14 International Class-H records.)

War shortened the 1914 racing season, as it was to do in 1939, but for the Easter Brooklands Meeting, Dario Resta (white coat and sleeves, navy blue cap with white stars) drove the Sunbeam in the First Lightning Short Handicap race. He started from scratch in company with L.G. Hornsted's pre-war 21,504 cc Benz and came home in second place to N. Holder's 4.5-litre Vauxhall which had given Resta a start of 32 seconds. The Sunbeam lapped at 113.45 m.p.h., and the big Benz was third, having lapped at 112.42 m.p.h. Resta non-started in the Lightning Long race. 'Toodles V' did not appear at the Track until the fateful August Bank Holiday meeting, when Resta won the Lightning Long Handicap in the bright aluminium-hued Sunbeam from scratch at 106.88 m.p.h., having lapped at 111.92 and 113.97 m.p.h. A re-handicap of 'owes three seconds' was too much for Resta in the Lightning Short race. Previously the car had competed at Saltburn sands, where Resta won his race, and over the f.s. kilo clocked 117.6 m.p.h. in one direction, averaging 111.05 m.p.h. Not bad on a beach course.

The war having ended motor racing in Europe, the Sunbeam was shipped to still-neutral America, where Ralph de Palma won a Match Race against a Peugeot and a Blitzen Benz at the then-new Sheepshead Bay, Long Island track. It apparently then became involved in a serious accident at Kalamazoo dirt track, when it skidded across the course and eight cars collided with it, causing ten deaths, it is said. It seems that the remains of 'Toodles V' were obtained by the Packard Motor Co. and

The 9-litre V12 Sunbeam 'Toodles V' being refuelled on the Paddock entrance road at Brooklands, prior to its attempt in 1913 on the World hour record.

after examining it they designed what was to become the first production 12-cylinder car, the Packard Twin-Six. The war had robbed the Sunbeam Motor Car Company of the chance to achieve this milestone, according to a guarded statement made by Louis Coatalen at a Board meeting.

However, the Sunbeam-Coatalen Mohawk aero-engine served with distinction from 1915 to 1917 in RNAS airships, having been developed to give 225 hp.

2. The 1919 350 hp V12 Sunbeam

Louis Coatalen put this V12 Sunbeam in hand in 1919, before Brooklands had been repaired following the ravages of war. The basis of this great car was a Sunbeam 'Manitou' engine of the kind which had served the RNAS creditably during the war in coastal patrol seaplanes. In this form this was a 48-valve 110×135 mm. engine using four six-cylinder magnetos, which weighed 845 lb dry and was rated at 300 hp at 2,100 r.p.m.

For the racing car Coatalen used a rather different version of this engine, the bore being 10 mm. greater and each cylinder having three instead of four o.h. valves, one inlet and two exhausts, per cylinder. Whether this 18,322 cc engine was a later military version of the 'Manitou' or whether Coatalen had it modified specially for the car, I do not know, but it is interesting that Ettore Bugatti changed from four to three valves per cylinder for his wartime aeroplane engine designs. Certainly, contemporary accounts go to some length to emphasize that the engine was built specially for the car, but then they also refer to it as of 400 hp. In fact, as tuned by 'Bill' Perkins, the ultimate power output at Brooklands was 355 b.h.p. at rather more than 2,000 r.p.m., with a b.m.e.p. of 118 lb, at 1,750 r.p.m. Post-Armistice Sunbeam aero engines, incidentally, of greater power than the 'Manitou', used three valves per cylinder, but operated by a camshaft between the blocks instead of by o.h. camshafts.

The engine of the big Sunbeam is a complicated and fascinating unit even by today's standards. The cylinder blocks, each containing three cylinders, were cast in light alloy with integral inlet manifolds and bolted to an alloy crankcase in two banks to form a 60-deg. 'vee'. Light-alloy water plates were attached by rows of 2 b.a. screws to the sides of these blocks and the cylinder heads were integral. In the car engine the four valves per cylinder of the aero-engine gave place to two exhaust and one inlet valve, each with two valve springs, operated by one o.h. camshaft above each bank of cylinders. The camshafts were driven by a train of spur gears from the front of the crankshaft and the timing case had a platform cast with it to accommodate the magnetos. Four magneto drives were provided, as on the aero-engine, but of these only the inside ones were used, two B.T.H. twelve-cylinder magnetos firing two K.L.G. plugs in each cylinder through 120 ft. of h.t. lead. The crankshaft ran in eight bearings, and there were two connecting-rods on each crankpin.

The induction arrangements were most imposing. A broad 'T'-shaped, water-jacketed manifold coupled each opposite pair of cylinder blocks, these transverse manifolds seating on the upper faces of the inlet manifolds. To the under surface of each 'T'-manifold was bolted a twin-choke H.C.7 Mark II Claudel-Hobson 42 mm. carburettor, with a barrel-throttle of some 2–$2\frac{1}{8}$ in. diameter. To the air intake of each carburettor was attached a 4 in. diameter pipe which turned through 180 deg. and rose, in front of the 'T'-manifold, to above the centre of the 'vee' formed by the cylinder blocks. The exhaust ports were on the outside of each block, long exhaust pipes running down each side of the car.

Cooling was by a single water-pump, gear-driven from the front of the engine, each cylinder block having two copper water off-take pipes, or eight in all. On top of the engine alone there were no fewer than 48 hose clips. The honeycomb radiator had a brass shell and was trunnion-mounted. Lubrication was on the dry-sump system, two pumps being situated at the front of the crankcase. Oil was contained in the first twelve inches of the car's tail and was delivered to the engine and returned from it via 1 in. diameter pipes. A drive for the rev-counter was taken from the front of the nearside camshaft. The exposed flywheel measured approximately 22 in. in diameter. This engine, tuned by 'Bill' Perkins, gave 355 b.h.p. at 2,100 r.p.m. and a b.m.e.p. of 118 lb. at 1,750 r.p.m., and was mounted in an H-section, three-point-suspended sub-frame.

The chassis in which Coatalen installed this impressive engine was a conventional channel-section frame, having a maximum depth of 4.75 in. and a maximum width of 29.5 in. The wheelbase was 10 ft. 7 in., the track 4 ft. 6 in., the car being 3 ft. 10 in. high to the top of the radiator cap and 16 ft. in length. A dry-plate clutch was used, actuated by a pedal on the nearside of the cockpit, this pedal having a fairly long travel. A clutch-stop in the form of a small contracting band-brake was an interesting

The famous 350 hp V12 Sunbeam as it first emerged from the racing department, with artillery wheels and a temporary windscreen.

feature. The four-speed gearbox was mounted on two cross tubes and operated by a right-hand lever inside the body. It has been rumoured that the car had no reverse gear, but the gate certainly has a reverse position. An open propeller-shaft with two universal joints conveyed the drive to a bevel-drive rear axle giving a top-gear ratio of 1.5 to 1, this axle having a steel casing turned from the solid. Suspension was by cord-bound half-elliptic springs, shackled at the rear, torque being absorbed solely by the rear springs. Dual Hartford shock-absorbers were fitted, and the front axle was of normal I-section, machined from the solid, and having Elliott-type steering heads. Wire bracing ran from the front cross-member to the centre of the main cross-member, beneath the radiator. The brakes consisted of an 18-in. diameter ribbed drum behind the gear-box, the shoes of which were connected by cable to a hand-lever outside the body on the offside, and large brakes on the rear wheels, operated via cables passing over brass pulleys at the extremities of the pedal cross-shaft. The steering box was mounted on the rear of the engine, the cord-bound centralized wheel operating the drag-link, on the offside, through a steel universal joint. Many of the chassis parts were cut off from the solid.

A very slim single-seater body was fitted, the short, streamlined tail being formed as a combined oil and fuel tank mounted on the rear chassis cross-member, the front of this tail being carried by brackets bolted to the frame, and a strap passing over the tail. A curious big-bore vent pipe about 12 in. long extended vertically from this tank. The radiator was enclosed in a conventional cowl, this and the undershield, chassis frame and tail being painted Sunbeam grey-green, while the long

The 350 hp Sunbeam in the form in which it was raced, with wire wheels.

bonnet, scuttle and cockpit sides were of polished aluminium. The air-pump for the fuel feed was on the nearside of the cockpit and the instrument board carried, from left to right, two magneto switches, an oil-pressure gauge reading to 18 lb./sq. in., a small speedometer (the present one reads to 180 m.p.h.), water thermometer, a protruding rev-counter reading to 2,600 r.p.m., a Sunbeam pressure gauge marked 'Full Load-35', and an air-pressure gauge. Ignition and mixture controls were above the steering wheel. Originally artillery wheels were fitted, but when the car emerged from the Wolverhampton works Rudge wire wheels carried 880×120 covers. The dry weight of the car was 28 cwt.

The Sunbeam was completed and fitted with its narrow single-seater racing body, undershield and cowl over its radiator, just in time for the Brooklands Whitsun meeting of 1920. There it was to have raced against the big Benz, but Harry Hawker, the famous airman, who had agreed to drive it, elected to do a practice lap on the Monday morning. The offside front tyre burst on the Members' banking and Hawker finally lost control on the Railway Straight, although the deflated tyre appears to have stayed on the wheel. The big car plunged through the corrugated-iron fence on the railway side, damaging its radiator and steering. Hawker was uninjured but it was not possible to race the car. In fact, it was Sunbeam's unlucky day, because that afternoon a six-cylinder Sunbeam left the track within sight of Hawker's accident and rolled over near the sewage farm. It is thought that Harry Hawker may have had another excursion off the track, apart from his trip through the Railway Straight fencing, and it seems that following the latter mishap the differential was deleted from the Sunbeam's back axle.

Mishap! The 350 hp Sunbeam after a tyre had burst in practice and taken Harry Hawker through the fence bordering the Railway Straight at Brooklands in 1920, without serious injury to driver or car.

The V12 Sunbeam was next billed to appear at the Brooklands August meeting, Hawker being entered for the Sixth Lightning Short Handicap. Alas, the driver stalled the engine on the line and the car had to be pushed to the side of the track.

The Sunbeam was not entered for the Autumn meeting because it was due to be shipped from Southampton to France to compete in the Gaillon Hill Climb, where René Thomas was to drive it. This hill climb was up a kilometre course with an average gradient of about 9% on the main road near Paris. There were no corners and competitors received another kilometre in which to work up speed before being timed, but a corner reduced this to virtually a 600 yd. flying start. The difficulty was in the state of the road, which military transport had made very rough, so that it was not easy to hold a very fast car to its course. The French were not happy that a German Benz had held the record since 1913, and Thomas had already made an unofficial attack on this 'hun' record with an Indianapolis Ballot, and had succeeded. The Sunbeam, which was being credited with a maximum speed of 175 m.p.h. if its engine could reach 3,000 r.p.m.(!), was now fitted with wire in place of artillery wheels.

Gaillon competitors were required to carry a passenger. This Thomas couldn't do, as the Sunbeam was a single-seater. He was allowed to carry lead ballast, bolted to the floor near his feet. The car ran with unsuitable gear ratios—I suspect exactly as it had run at Brooklands; it certainly had the same size Palmer cord tyres—the only preparation being the fitting of extra pairs of Hartford shock absorbers, four per axle. Thomas made two practice runs, and on the day handsomely broke Earle's record with the Benz, climbing in 20.6 seconds (108.3 m.p.h.), whereas the German had clocked 23.0 seconds. Thomas had twice had his foot knocked from the accelerator when the ballast, which had broken away, lifted from the floorboards over the bumps in the road. One item of equipment of the Sunbeam I find extremely pleasing—a bulb horn carried on the offside of the scuttle!

After this rousing performance at Gaillon the car was rested for the remainder of 1920.

René Thomas was invited to drive it at the 1921 Brooklands Easter meeting. However, it was Kenelm Lee Guinness, the first man to make the big car perform properly, who took it out for the start of the 8th Lightning Short Handicap. Again Brooklands proved unlucky for the big Sunbeam, for, getting away from the scratch mark—'Chitty-Bang-Bang' had left 12 seconds earlier—second gear broke up and the car rolled to a standstill. It started in the Lightning Long Handicap when Guinness, hampered by loss of second speed, eventually got going really well, to finish behind Zborowski's winning 1914 GP Mercedes, which had received a handicap start of 39 seconds. Guinness did one lap at 119.15 m.p.h., and the next at 120.01 m.p.h.

It was never easy for Coatalen to find drivers sufficiently brave and experienced to handle the Sunbeam. For the Whitsun Brooklands races he delegated the task to L.G. Hornsted, but he failed to get much speed from the car, his fastest lap being at only 109.7 m.p.h., although he may have held back deliberately to fox Mr Ebblewhite, the Brooklands handicapper, who had set the Sunbeam to lap at 117.19 m.p.h. in the two Lightning Handicaps for which it was entered.

At the 1921 Autumn meeting the big black-and-white Sunbeam came

K. Lee Guinness, who became the best-known and most successful of the Brooklands drivers of the 350 hp Sunbeam, attacking standing-start records, front wheels on the pneumatic timing strip, the back wheels spinning. Note that the driver is wearing a collar and tie and is bare-headed. Bill Perkins, in cap, who looked after the car, is watching the getaway.

into its own and began to justify Coatalen's financial expenditure on it. In the hands of K. Lee Guinness, who drove in his customary skull-tight helmet, the Sunbeam ran home third behind Cook's Vauxhall (42 sec. start) and Hawkes's Lorraine-Dietrich (22 sec. start) in the 12th Lightning Short Handicap, its fastest lap at 114.49 m.p.h. It then came out for the 12th Lightning Long Handicap and fought a prodigious battle against Swain's red 1914 GP Vauxhall, which had been released 27 sec. before Guinness left from scratch. The Sunbeam tore after the field, averaging 102.9 m.p.h. to the Fork from the Pond start, completed its second lap at 116.09 m.p.h., and flashed across the finishing line beside the Vauxhall, which was declared the winner by the 'thickness of its radiator'. Guinness had been timed unofficially to cover the Railway Straight half-mile at 135 m.p.h. each lap and was said to have entered the finishing straight at 140 m.p.h. His last lap was done at 116.36 m.p.h., and his race average was 111.59 m.p.h. compared to the 101.34 m.p.h. of the victorious Vauxhall. Yet on this occasion the Wolverhampton monster was reported to be out of tune! The B.A.R.C. awarded 'K.L.G.' a special cup to commemorate this thrilling race.

The week following, the French driver Jean Chassagne tried the Sunbeam. He won the 13th Lightning Short Handicap by a length, from scratch, against the Lorraine and Rampon's ex-Duff Fiat—veritably a battle of the giants!—at 103.75 m.p.h., lapping at 111.92 m.p.h. This resulted in a re-handicap to 'owes four seconds' in the 13th Lightning Long Handicap, and Chassagne didn't appear to try very hard, slowing down, failing to lap faster than 114.49 m.p.h. and retiring before the finish when a tyre threw a tread.

How the spectators saw it. The big Sunbeam coming up the Finishing Straight at Brooklands, with the finish line and the Judge's Box just ahead of it.

By the May meeting even more performance had been extracted from the V12 engine, for when Guinness drove it (in place of Albert Divo) in the 14th Lightning Short Handicap he accelerated at an average of 105.43 m.p.h. to the Fork (the previous best had been 98.2 m.p.h.) and went on to lap at 118.86 m.p.h., winning at 111.42 m.p.h. from Rampon's Fiat, which had left 36 seconds earlier in this 5.75-mile race, and that in spite of being badly baulked on the banking, which necessitated passing below a slower car. Guinness started in the 14th Lightning Long Handicap and now averaged 109.12 m.p.h. to the Fork. He completed his second lap at 119.43 m.p.h., the next lap at 121.47 m.p.h., but after this meteoric performance the Sunbeam burst its offside front tyre coming off the Byfleet banking at 120 m.p.h., Guinness having a hard task to avert disaster.

It was obvious that the Sunbeam was now going really fast, and on 24 May Guinness went to Brooklands for a record attempt. On the previous evening he had been timed, by friends using car headlamps, to have covered the half-mile at about 144 m.p.h. The Wednesday turned out very windy, and after Perkins had tuned up the great car Guinness waited until 5 p.m. for the wind to drop. He then set out in the reverse—or clockwise—direction of the track, covering the half-mile at 140.51 m.p.h., the highest speed recorded at Brooklands up to that time, the mile at 127.88 m.p.h. and two miles at 122.11 m.p.h. Driving the normal way round Brooklands against the wind, he was timed to do 131.86, 130.48 and 125.61 m.p.h., respectively. This beat handsomely records held by the big Benz. As the Palmer tyres were unworn, Guinness went on to attack standing start records, spinning the back wheels in clouds of

Lee Guinness at the Track, with the aeroplane sheds behind the big Sunbeam.

rubber-smoke as he accelerated away. He averaged 76.85 m.p.h. against the wind, 76.59 with the wind, over the half-mile; 94.78 against and 98.54 m.p.h. with the wind over the mile. Over a kilometre the car was timed at a mean speed of 134.65 m.p.h. During the record attempts the Sunbeam set a new lap record of 123.39 m.p.h., this record not having to be made during a race in those days.

One-direction runs sufficient to break the Brooklands Class J records and World's honours were claimed for the s.s. half-mile (76.72 m.p.h.), s.s. kilo (134.65 m.p.h.), s.s. mile (96.63 m.p.h.), f.s. half-mile (136.05 m.p.h.), f.s. kilo (215.25 k.p.h.), f.s. mile (125.17 m.p.h.) and f.s. two miles (122.11 m.p.h.). The f.s. kilo record of 133.75 m.p.h. ranked as the World Land Speed Record, the last time this was broken on Brooklands track, endorsing the 350 hp Sunbeam as the fastest car in the world.

Incidentally, about this time S.C.H. Davis of *The Autocar* had a go in the car, lapping at 110 m.p.h.

On the Saturday following these sensational records, the Essex M.C. put on its Royal Brooklands meeting, attended by HRH the Duke of York, and Guinness proudly brought out the famous car for the Earl of Athlone Lightning Handicap. Alas, the offside rear tyre gave trouble when the car was travelling very fast, and it was left to Segrave with the 5-litre six-cylinder car to uphold Sunbeam honour that day.

Guinness and the V12 Sunbeam were entered for the Whitsun races, but failed to start, and the Sunbeam's last appearance at Brooklands was on 30 September 1922 when Guinness won his heat and the final at the Essex M.C. Championship Meeting, gaining the over-5,000 cc Championship. In these 5.5-mile scratch races Guinness won the first, from Barlow's Benz, at 110 m.p.h., and the second, from the Leyland-Thomas, at 104.04 m.p.h.

After this the Sunbeam languished at Brooklands. Incidentally, Coatalen had promised in 1920 that it would return to Gaillon properly geared for hill climbing, but this promise was never fulilled and the highest speed attained at the 1921 hill climb was 90 m.p.h., by a Ballot. Malcolm Campbell was extremely anxious to own the fastest car in the world, and he persuaded Coatalen to lend him the car to drive in the Saltburn Speed Trials of 1923. Here he made six runs, clocking 138 m.p.h. for the best run in one direction, over the kilometre, and averaging 134.07 m.p.h. for the best two-way runs. Campbell returned home confident that he had beaten Guinness's World record, but the runs had been hand-timed and were not recognized.

Determined to put his name amongst the long list of Land Speed Record holders, Campbell bought the Sunbeam from Coatalen at a price which has never been disclosed. The car was sent by rail to Campbell's home at Povey Cross, being driven on the road from Horley Station. It was found to be in poor condition, the oil scavenge pump out of action and the gearbox pinions stripped. Although the Sunbeam Company offered every assistance it seems probable that Coatalen was aware that the Sunbeam was nearing the end of its useful career and, as it was really too fast for Brooklands on the tyres then available, he didn't mind disposing of it.

Hectic work by Campbell and his mechanics under Leo Villa, assisted by a mechanic, Webster, on loan from Wolverhampton, got the car ready for the Danish A.C. Fanoe Island Speed Trials in 1923. Even so, Webster, with a new gearbox shaft under his arm, only just caught the train as it was moving out of Liverpool Street. Further long hours of toil were necessary on arrival at Fanoe, especially as on the only practice run the shock absorber brackets broke. In the event the Sunbeam won a scratch

The re-bodied Sunbeam as used by Capt. (later Sir) Malcolm Campbell, to take the Land Speed Record to over 150 m.p.h. for the first time, on Pendine sands in S. Wales.

race against a big Opel and aero-engined Stoewer, at 82.19 m.p.h., and clocked a mean speed of 137.72 m.p.h. over the two-way mile, and 136.32 m.p.h. over the two-way kilometre—its fastest speed over the mile being at 146.4 m.p.h. Once more, however, the International authorities refused to recognize this speed as a World record, although the runs had been electrically timed by apparatus tested and certified as accurate in Paris, and supervised by Ebblewhite.

Campbell returned disgruntled but not defeated. He set Villa to extract more power from the old V12 engine and got Boulton & Paul of Norwich to build a new body for the Sunbeam. This actually consisted of a long tapering tail with head fairing mounted over the Brooklands petrol tank. The cockpit was made wider to mate up with this new tail, and to raise the height of the scuttle a new top section was welded in. A longer radiator cowl was made up, the rear springs faired in, the external hand-brake given an aerofoil fairing and Ace discs fitted to the wheels—usually only on the back wheels, while the aero-screen was dispensed with. The Sunbeam was repainted blue and the name Sunbeam removed from the bonnet, but this particular Malcolm Campbell car was not usually called 'Bluebird'. Stub exhausts sometimes replaced the long tail pipes.

Campbell took the car to Fanoe beach in 1924 but met with little success. On an early run two back tyres left the wheels, yet Campbell

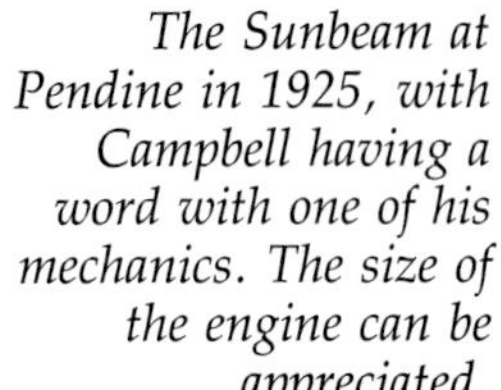

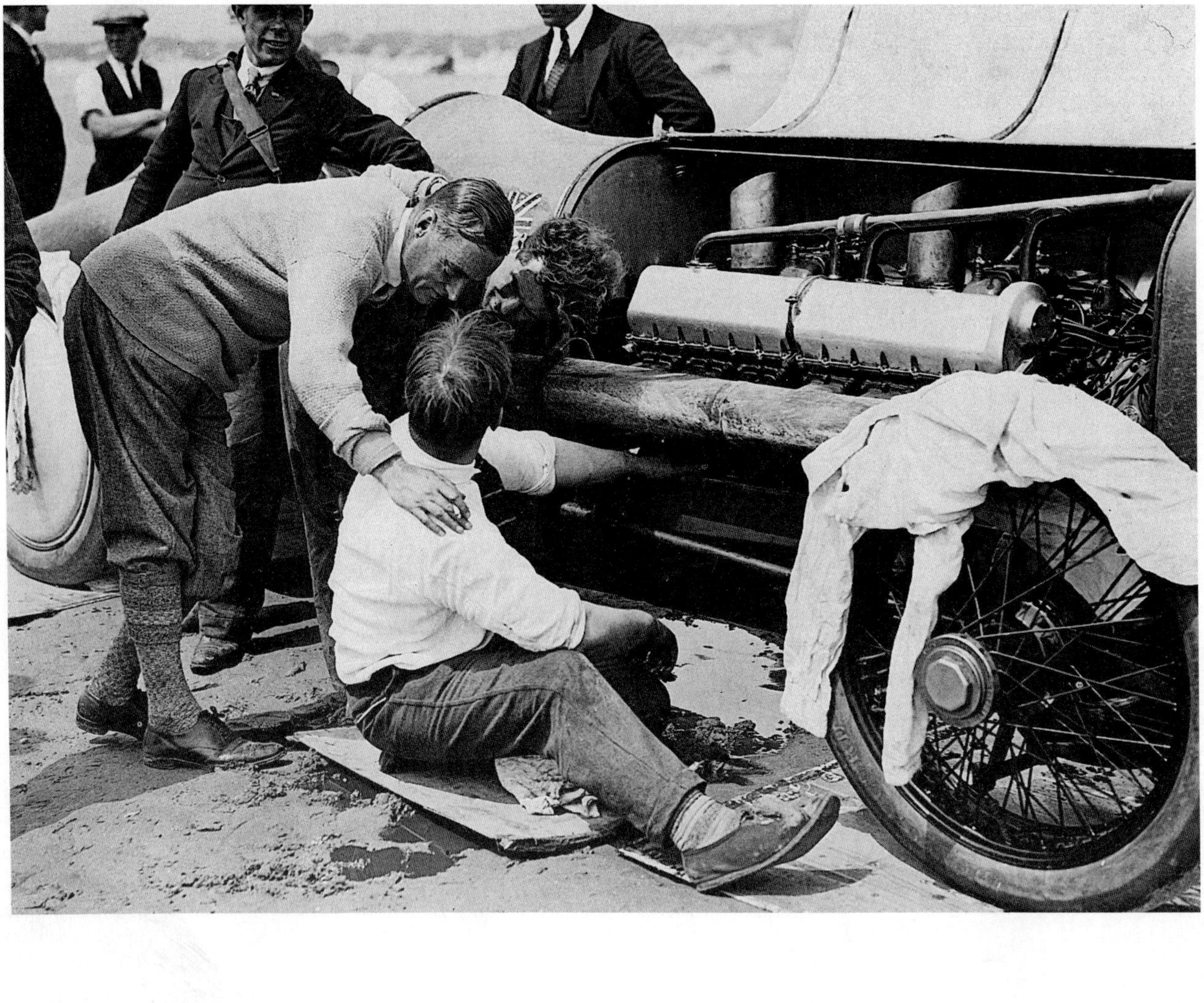

The Sunbeam at Pendine in 1925, with Campbell having a word with one of his mechanics. The size of the engine can be appreciated.

brought the car safely to rest. Straight-sided tyres were then fitted to the back wheels but beaded-edge tyres were retained on the front wheels. The offside tyre left the rim on a subsequent run and hit a boy spectator, who died from his injuries.

Back again to Povey Cross came the old Sunbeam, and new camshafts were machined for it in the workshops there, resulting in a further increase in engine power. Campbell considered taking the car to Southport or Saltburn sands for his next attack on the Land Speed Record. In September 1924 he journeyed to Pendine Sands in Carmarthenshire. Here he was at last successful, being timed to cover the two-way kilometre at 146.16 m.p.h. His mean time was a mere 0.015 of a second faster than that of Eldridge's 300 hp Fiat which had broken Guinness's record at Arpajon in France. After this Campbell advertised the Sunbeam for sale, at £1,500.

Campbell now put in hand a special car powered with a 450 hp Napier 'Lion' engine, but, aware that the Sunbeam Company had a new 4-litre supercharged V12, Thomas, a Liberty-engined monster, both destined to try to wrest his record from him, Campbell decided to try again with the old Sunbeam. He gave the car a preliminary outing at the 1925 Skegness Speed Trials in June, where the Sunbeam won the unlimited racing car event, clocking 33.4 seconds, but was 0.4 of a second slower than

A happy shot of Capt. (later Sir) Malcolm Campbell in the Sunbeam's cockpit. He continued the car's notable performance after its Brooklands days were over. Note the streamlined handbrake and the Union flag.

E.A. Mayner's little Mercedes, which made f.t.d. The Sunbeam had its Brooklands radiator cowl and exhaust on this occasion, and the back axle was not faired in, although the new tail was in place. On 21 July 1925, at Pendine, the old Sunbeam encountered better conditions, Campbell being timed at a mean speed over the two-way kilometre of 150.87 m.p.h., and over the two-way mile at 150.766 m.p.h. His best run was at over 152 m.p.h., and Campbell had achieved his ambition of being the first driver to exceed 150 m.p.h.

Just before this successful attempt Campbell had driven the Sunbeam in the Pendine Speed Trials, making f.t.d. by covering the standing start kilometre in 43.0 sec. (83.7 m.p.h.), winning the unlimited open and winner's handicap races. Even now the full streamlining had not been fitted to the car. To commemorate being the first driver to exceed 150 m.p.h., Campbell had some non-working big scale models made of the 350 h.p. Sunbeam, of which at least two are still in existence.

Losing his record the following year to Segrave and the 4-litre Sunbeam, Malcolm Campbell pressed on with his new Napier-Campbell, with which he embarked on a series of seven further successful onslaughts on the World Land Speed Record, culminating in a speed of 301.13 m.p.h. at Utah in 1935, and for which exploits he received a well-earned knighthood.

For some time the old V12 Sunbeam stood neglected in the Povey Cross workshops. It was bought by Ralph L. Aspden, who learnt to drive it on the road (!) prior to running it at Southport sands. The clutch proved difficult to disengage and Aspden rigged up an electric servo for it. He did not achieve much with the old car, which he sold to Jack Field.

Field made a few appearances with the car before selling it to Billy Cotton, the dance-band leader. He did little with it, but did do a timed run on Southport Sands, at 121.5 m.p.h., rewarded with a Southport 100 Gold Badge.

When war came in 1939 the famous Sunbeam disappeared. It was eventually tracked down in Lancashire by Harold Pratley, who had

The historic Sunbeam as Harold Pratley discovered it, abandoned, during the war. He saved it until it was acquired by Lord Montagu of Beaulieu.

heard that it was standing behind a garage on the Liverpool–Southport road. It is to his lasting credit that he prised it away from its new owner just in time to prevent it from going for scrap. Indeed, while Pratley was still negotiating, the Sunbeam, now in a sorry state, was towed through Liverpool behind a cart-horse to attract attention to a 'Books For War Salvage' drive! That was in 1943, and Pratley bought the car in 1944 and towed it to Derby, and after the war had it brought by rail to Walthamstow.

In 1946 the Sunbeam was lent to Rootes Ltd, who cleaned it up and painted it, so that it looked much as it had done in its Brooklands days, albeit the name on the bonnet wasn't quite correct and the Campbell cockpit remained. The Sunbeam was taken to Belfast on a Commer truck for the S.M.M.T. Jubilee Cavalcade. Two years later it was exhibited at the Victoria League Racing and Sports Car Exhibition at Henley Hall, London, and it took part, on tow, in a Woodford Road Safety Parade and was displayed in the showroom of a local Vauxhall dealer.

Pratley had come to the conclusion that the big Sunbeam would never run again. In 1956 he told me that most of the alloy castings had corroded, the many 2 b.a. studs securing the water jacket plates had rusted in, the engine was solid and not even the valve springs could be depressed. In addition, water pipes fell apart on being dismantled, the

Lord Montagu of Beaulieu racing the old 350 hp Sunbeam at Oulton Park, after the war.

base of the engine was full of sand and the whole engine caked with 'R' oil that had gone solid. Only the K.L.G. plugs looked new! Rebuilding would involve stripping the engine, replacing scores of water hoses, drilling out and re-tapping masses of screws, replacing 120 ft. of h.t. lead, 36 valves, 72 valve springs, cleaning yards of oil-ways, etc.

However, the Montagu Motor Museum took this famous racing giant in hand in 1958. It was meticulously restored but did not immediately become a static museum exhibit. Indeed, the old Sunbeam was billed to appear at a demonstration of historic racing cars at a B.R.D.C. race meeting at Silverstone in May 1959. Unfortunately, this was not possible, because the gearbox broke up beforehand—a recurring problem! The Montagu Motor Museum's engineer, Mr Warne, put in much hard work in order to have the Sunbeam ready for Lord Montagu to drive it at that year's V.S.C.C. Oulton Park race meeting. A vintage Albion lorry gearbox had been installed, and His Lordship, whom I know to be a very accomplished driver, paid tribute to Warne's hard work and drove the Sunbeam notably fast. Warne, for his part, paid tribute to Edward Montagu's skill, saying that the power of the big engine comes in suddenly at the top end of the rev. range, making the car difficult to control. Alas, earlier in 1959 Bill Perkins, the dedicated mechanic who administered to the car in its Wolverhampton days, had died.

From Oulton Park the Sunbeam went to Blackpool, on display, and at the B.A.R.C. Festival of Motoring at Goodwood in July 1962 it was demonstrated in the company of 'Bluebird', the L.S.R. car of Donald Campbell CBE, son of the Sunbeam's former driver, who closed the day

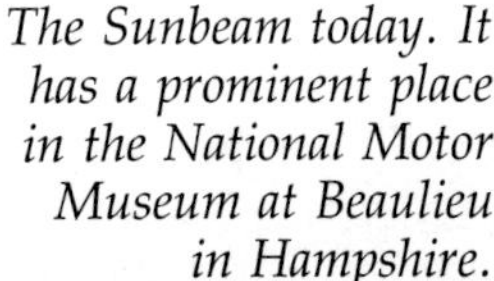

The Sunbeam today. It has a prominent place in the National Motor Museum at Beaulieu in Hampshire.

by driving the car. He described it as 'quite terrifying' and said that Sir Malcolm had regarded it as having the finest acceleration of any car he had driven.

Today the Sunbeam has a place of honour in the National Motor Museum at Beaulieu, on a dais with those other successful Land Speed Record cars, the 200 m.p.h. twin-engined Sunbeam, the Napier-engined 'Golden Arrow', and the aforesaid 'Bluebird'. It wears a wider radiator cowl than it had originally, the tyres are of bigger section, and that Albion gear-lever is longer. But how fortunate it is that such a celebrated and long-lived racing car has survived. Although at the time of writing it is a non-runner, a replica of the long tail used by Campbell at Pendine has been fitted in recent years, the original which Pratley had preserved being beyond repair; and the correct size wheels and tyres have replaced the oversized ones on which the car was raced beforehand.

3. The 'Chitty-Bang-Bangs'

'Chitty-Bang-Bang', that aeroplane-engined chain driven monster racing car which the late Count Zborowski introduced to the Brooklands' fraternity from the commencement of the 1921 season, has become a legend. The name, generally excused as denoting the noise made by the engine as it idled over, but in reality based on the theme of a lewd wartime song, a joke entirely in keeping with the mercurial temperament of the car's owner, has endured down the years, long after the car herself has passed into oblivion. It has become a household term for anything uncouth and enormous. Rendered originally as 'Chitti-Chitti-Bang-Bang', but later spelt with a 'y', there have even been children's fantasies of the name, written by none other than the late Ian Fleming, creator of the immortal James Bond, Secret Agent. Incidentally, when I express my dislike of comic nicknames for modern racing cars, a sagacious friend never fails to remind me that I find such names quite acceptable on Brooklands' racing cars. I can only add in deference that owners had to pay the Track authorities a fee to register them in those days, and that usually they only appeared in the programme after the make of car had been quoted. One of the exceptions being the giant that was the first of Zborowski's stud to be called 'Chitty', which was quoted in B.A.R.C. Race Cards as 'Chitty-Bang-Bang'.

In fact there were three 'Chittys', with a fourth on the drawing board when Zborowski was killed while driving a works Mercedes at Monza in 1924.

Brooklands after the Armistice was essentially a place where handicap racing was contested between a wide variety of motor cars. It was also, during those 'mad twenties', the haunt of amateur racing motorists, who greatly outnumbered the professional racing drivers. Then, as now, the more money there was available, the greater the chances of success.

To compete with distinction against potent factory-built racing cars such as the single-seater 350 hp V12 Sunbeam and the slim 16-valve A.C.s, not to overlook the better pre-1915 Grand Prix cars that had found their way to post-war Brooklands, represented the problem that confronted the less-affluent would-be motor racing aces.

Count Louis Vorow Zborowski, born of English and American parents, educated (briefly) at Eton, and residing at Higham, a gentleman's estate near Canterbury in Kent, was anything but impecunious. Obviously he built 'Chitty-Bang-Bang' for fun and from a sense of adventure, rather than as a financial expediency, because at the time he was racing her he could also afford to run the winning 1914 GP Mercedes and a 1919 straight-eight Ballot, etc., and was soon to race a Type 30 Bugatti at

Indianapolis and a Salmson in the first J.C.C. 200 miles race, import from the United States a 2-litre eight-cylinder Miller and invest heavily in the Aston Martin concern in order to enter the more exacting sphere of road racing.

The fact is that Count Zborowski's 'Chitty-Bang-Bang' was the first of the amateur-built post-war aero-engined track monsters, and in its day it was the fastest of them, with the exception of the 350 hp Sunbeam.

The construction of 'Chitty I' was put in hand at the workshops at Higham and the car was completed by March 1921.

Captain (later Lt.-Col.) Clive Gallop, who had served his apprenticeship with Peugeot in France before the war, acted as engineering consultant as the work proceeded in the Higham workshops. Zborowski obtained his engine—a six cylinder 165×180 mm. (23,093 cc) Maybach aeroplane motor—from the Disposals Board; it had four overhead valves per cylinder operated by exposed push-rods and rockers. Nominally it was of 300 hp. In fact, it developed fractionally more than this, at the decidedly modest crankshaft speed of 1,500 r.p.m. and was the type of engine which had powered German Gotha bombers.

The chassis into which this enormous engine was coaxed was a Mercedes, usually quoted as a pre-1914 75 hp type but lengthened, I believe, for the purpose. The Mercedes radiator, scroll clutch and gearbox were retained, final drive being by side chains, but the axles were modified to take Rudge-Whitworth hubs and wire wheels shod with 895×135 Palmer Cord tyres. In order to get sufficient ground clearance the original Maybach oil sump had to be removed, so a system of dry-sump lubrication was contrived, with a big oil reservoir slung alongside the chassis frame on the offside. In the fashion of those times this was given a pointed prow for good air penetration, in spite of the enormous

The immortal 'Chitty-Bang-Bang I' in its original form, outside the house Higham, at Bridge near Canterbury in Kent, home of its creator Count Louis Vorow Zborowski, where it was built in the well-equipped workshops.

drag of the big radiator and primitive bodywork.

'Chitty' was in the nature of an experiment and it was at first given a crude four-seater body, which was knocked up by Bligh Bros. of Canterbury, a very long-established coachbuilding firm in which the Count had a financial interest and whose trade plates he apparently used when taking his cars out on test. 'Chitty' was an immediate success and was great fun on the road, providing its negligible braking power was kept in mind, and it was not called upon to negotiate twisting sections of the local terrain.

At Brooklands, the great car was a sensation. Zborowski used to turn up in company with his friends Major 'Shuggar' Cooper, Jack Cooper, Miles and mechanic Wigglesworth, all of them wearing loud check caps imported from Palm Beach. To see them starting 'Chitty' with the aid of a half-axle from a B.E. aeroplane, the half-compression device, and someone furiously winding the dashboard starting-magneto was a sight no Brooklands' *habitué* cared to miss.

The Press helped the drama along by describing 'Chitty' as of 600 hp and billing it as a rival to the big Sunbeam, which they endowed with at least 450 hp. For all his flamboyance, Zborowski was cautious. He knew that much would be expected of his 23-litre home-built car and that if it failed as a racing car he could become a laughing stock. He did not expect

'Chitty-Bang-Bang I' being backed out of the garage at Higham by the Count. The exhaust-pipe extension was a bit of fun on the part of Chitty's crew! The tyres are Palmer Cords, very popular for racing in the early 1920s.

the handicappers to treat him kindly. So the ugly body, set off by a crude exhaust pipe with a right-angle bend in it, was a foil to any ambitions he had for the car.

Painted grey, 'Chitty' was entered for the Easter 1921 Brooklands meeting. Although the chassis had been strengthened by bracing rods and flitch plates, handling was none too good and it was found advisable to carry 7 cwt. of sand in the back of the body to keep the back wheels in contact with the cement.

'Chitty' went out for the first race on that Easter Monday at Brooklands on 28 March 1921, giving Birkin's D.F.P. 78 seconds start, and itself leaving the line four seconds after André Boillot in the 4.9-litre six-cylinder Sunbeam. Zborowski was faster on his standing-start lap than the Frenchman, and although the Sunbeam made up a little on the flying lap it could not catch the aero-engined car, which lapped at 108.15 m.p.h. and won its first race, this 23rd 100 m.p.h. Short Handicap, at 100.75 m.p.h. The 'bookies' took 6 to 4 on Zborowski, 4 to 1 the field.

The car came out again for the Lightning Short Handicap. It started 12 seconds after the favourite, K. Lee Guinness in the V12 Sunbeam. Unfortunately, as the Sunbeam's clutch was engaged the gearbox broke up, so 'Chitty' had a walk-over, winning at 100.5 m.p.h. from Duff's 10-litre Fiat, a mere baby, while Zborowski's 1914 G.P. Mercedes driven by

The method of starting the 23-litre Maybach engine of 'Chitty I' was to first pull it over with a bar made from part of an aeroplane's axle.

Ready to race. Zborowski in the revised 'Chitty-Bang-Bang I' with pointed tail and cowled radiator. He and his enormous motor car are an enormous attraction for the spectators in the Brooklands' Paddock.

Hartshorne Cooper finished third. 'Chitty' missed the 100 m.p.h. Long Handicap, as a broken petrol pipe had to be repaired, but it was out again for the Senior Sprint race in which, heavily re-handicapped, it was content with second place behind Zborowski's 4.5-litre G.P. Mercedes, again driven by his friend. As the Count had won the Lightning Long Handicap in this car in the meantime, his Easter egg must have tasted very sweet!

His driving of the big car on its first public appearance had been warmly praised and the car had behaved splendidly, apart from some clutch slip when the chassis flexed over the notorious Brooklands' bumps. The standing laps were fast, in spite of the care and skill demanded by the scroll clutch, which had to be engaged carefully with the engine idling, to protect the transmission, and the push-start needed before the second race. The Sprint race might have been won had not the other Mercedes got somewhat in 'Chitty's' path. Speeds of around 115 m.p.h. over the half-mile had not resulted in tyre trouble.

All told, the Count must have looked forward to the Whitsun races. For them 'Chitty' had been somewhat cleaned up. The two Zenith carburettors, one at each end of a common inlet manifold, had been replaced by three Claudel-Hobsons, each with its own two-branch manifold, and the crude exhaust system had been changed for a six-branch manifold and outside pipe. Moreover, the 'touring' body had been removed and a two-seater with duck's back tail substituted, painted black. Guards had been fitted above the driving chains. The confidence Zborowski must have had in 'Chitty' was marred by the fatal accident in practice which befell his friend Hartshorne Cooper while testing an aero-engined car he had built on the lines of 'Chitty-Bang-Bang'. However, this did not deter the Count from racing on the Whit-Monday.

Hornsted on the V12 Sunbeam gave 'Chitty' ten seconds start in the Lightning Short Handicap and Zborowski never saw the Wolverhampton car. He lapped at 111.92 m.p.h. and won easily, averaging 101.6 m.p.h. A re-handicap proved too much for 'Chitty-Bang-Bang' in the next race, in spite of a lap speed of over 113 m.p.h. and something approaching 120 m.p.h. down the straight. A cowl had been fitted over the radiator but fell off. The petrol tank also split as the flexible chassis rode the Brooklands' bumps, and could not be repaired in time for the sprint race.

The combination of Zborowski and 'Chitty' continued to pile up successes, however. At the 1921 Brooklands' Summer meeting they left from the scratch mark in the Lightning Short Handicap, taking second place behind Swain's G.P. Vauxhall, but were handicapped out of the Lightning Long race. Chitty had now been fitted with four Hartford shock-absorbers to keep the back wheels down, and went round the track with a curious lolloping ride.

It was not only by its track performances that Zborowski judged 'Chitty'. He greatly enjoyed driving the monster on the deserted roads of the times, and by the summer of 1921 had put in hand an improved version, known as 'Chitty II'.

This was on the same general lines as the first 'Chitty' but with a shorter wheelbase. The engine was a six-cylinder overhead valve Benz BZ IV series aero-motor of 145×190 mm. (18,882 cc), rated at 230 nominal horse-power. The chassis was again a pre-war chain-drive Mercedes, a 75 hp or Sixty model presumably, in which case the frames of both these 'Chittys' dated back probably as far as 1907, or even earlier. 'Chitty II' was given a narrow four-seater body, again the work of the coach-and-four people in Canterbury and, indeed, had been conceived more as an exciting road car than as a racing car. It was provided with excellent 'period' mudguards, lighting equipment and a hood and screen, and proved an excellent touring vehicle, providing its lack of retardation from the rear-wheel and transmission brakes was borne in mind, a point the effervescent temperament of the Count was apt to overlook, so that his close friends had very real fears for his life.

It had been Zborowski's intention to enter both 'Chittys' for the Brooklands' August Bank Holiday meeting and 'Chitty II' was mentioned in the preliminary entry lists, albeit with its engine dimensions incorrectly quoted. However, it was not ready and neither car enlivened this meeting.

At the Brooklands' Autumn meeting both 'Chittys' ran, but good fortune eluded them. The smaller car, painted black like the bigger one, proved under-geared for the Track and had too severe a handicap to overcome, being asked to give Major Segrave's straight-eight 3-litre Sunbeam eight seconds start in the 100 m.p.h. Short Handicap, and 12 seconds in the 8.5-mile version of this race. It lapped at 108.27 m.p.h. but was never again raced at Brooklands. 'Chitty I' was also off-form and finally broke a valve. Nevertheless, Zborowski had every reason to be pleased with his first season's excursion into the realms of fantastic motor cars.

Meanwhile, 'Chitty II' was earning her keep as a touring car; during the last week of January 1922 she was driven across France to Algeria, with the luggage following on in the White Mercedes. The Count and

Countess and their party not only had a memorable stay at the Negresco Hotel in Nice, but drove, on the spur of the moment, some 250 miles south of Biskra on the fringe of the Sahara Desert, 'Chitty's' radiator boiling dry, which could have ended in disaster had they been further in and used up all their reserve water supplies. Before 'Chitty II' was driven home to England she went up the east frontier of France to Strasbourg, so that Zborowski and Gallop could look at the Grand Prix circuit there, as they were to pit their 1.5-litre Aston Martins against the 2-litre cars in the French Grand Prix at that venue later in the year.

Although he was taking up Grand Prix racing and was involved with his Ballot when the 1922 Brooklands season commenced, Zborowski did not forsake the really big cars. He bravely drove Major Cooper's 200 hp Benz, a car dangerous to handle and thought to be in an advanced state of metal fatigue. I suspect that the broken valve had seriously damaged 'Chitty's' engine, or they may have been deliberately resting her to improve the car's handicap.

However, Zborowski substituted 'Chitty' for his Ballot in the Lightning Short Handicap at the Whitsun meeting, winning at 105.5 m.p.h. after passing Parry Thomas's Leyland-Thomas on the inside and crossing the finishing line with the offside back tyre in ribbons. It was during this

The grave at Burton Lazars, near Melton Mowbray, where the Zborowski family is buried. The Count had raced many other cars besides the 'Chittys' and was killed while driving a 2-litre GP Mercedes in the 1924 Italian Grand Prix.

The gravestone in memory of the young Count Zborowski.

The Conan Doyles, sons of the creator of Sherlock Holmes, with 'Chitty-Bang-Bang I' in a speed trial at Branches Park.

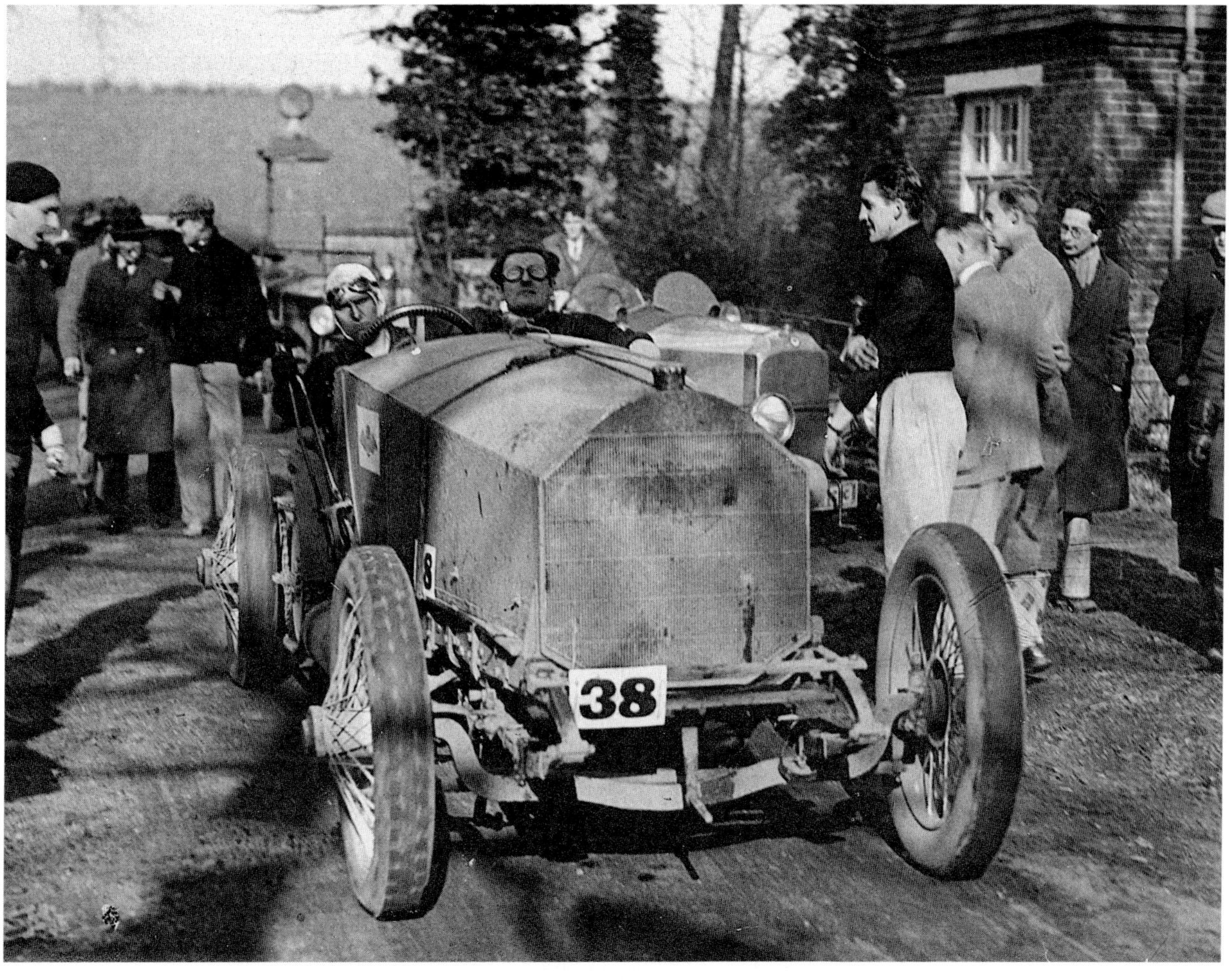

exciting race that 'Chitty' achieved her fastest-ever Brooklands' lap, 113.45 m.p.h. for the 2.75-mile circuit. Lack of tyres kept it away from the August racing.

The car came out again towards the end of 1922, winning the unlimited racing car class at the Southsea Speed Trials, where it covered a kilometre in 30.6 seconds, a speed of 73.10 m.p.h., beating the Isotta-Maybach with its T-head 180 hp engine and Barlow's Blitzen Benz. It also went very well at the Royal Brooklands' Meeting until the offside back tyre deflated and left the rim; its rival, the big Sunbeam, also having tyre trouble. For this race the original Maybach carburettors had been refitted.

Zborowski intended to resume Brooklands activities with 'Chitty' at the Essex M.C. Championship Meeting in September 1922. Alas, it was not to be! Lapping fast in practice the offside front tyre burst on the Members' banking and the car struck the parapet at the top of the track where the entrance road passed under it. 'Chitty' slid down the banking, turning round as she did so, went backwards through the wooden timing hut at the beginning of the mile, tore off her front axle and came to rest upright, a considerable distance away. The riding mechanic, Len Martin, was thrown out and badly shaken. Zborowski stayed in his seat and escaped serious injury. One official, Chamberlain, had the three fingers of one hand amputated as 'Chitty' smashed through the timing box he had been occupying. Another official, Cann, whose son became Gwenda Hawkes' chief mechanic, saw the car go out of control and slipped into the ditch, 'Chitty' passing over his head without doing him any harm. That was the last time the Count raced 'Chitty I' at Brooklands.

What happened to the 'Chittys' after his death? 'Chitty I' was acquired by the Conan Doyle brothers from J.E.P. Howey, who is said to have bought her from the Count before his fatal accident—they were partners in the Romney, Hythe and Dymchurch narrow-gauge railway project—to have painted her white, and used the old car for a tour of the Highlands of Scotland, afterwards racing her again at Brooklands at the 1924 Autumn meeting. 'Chitty' lapped at less than 97 m.p.h. before retiring, never to race again. They ran her in one speed trial, now in a coat of red paint, after which the historic car languished in a garage in London. At

'Chitty-Bang-Bang I' unsuccessfully awaited a buyer, after the Conan Doyles had tired of her and left her to languish at Brooklands Aerodrome.

Another shot of 'Chitty I' languishing at Brooklands in the mid-1930s, showing the three carburettors and their inlet manifolds.

my instigation she was brought out of retirement and displayed at a small exhibition of old racing cars in the Brooklands Paddock in 1934. R.G.J. Nash towed her most of the way, shedding tyres at all-too-frequent intervals, behind his Ford 'V8 coupe'. After the show 'Chitty' was put out to grass on the aerodrome. Her owners were hoping she might attract a buyer, but she was in a sorry state, the scroll clutch having lost its grip and the tyres their air. When the 200 hp Benz four-seater broke its gearbox the owner got hold of 'Chitty', hoping that the Mercedes gearbox might provide a substitute. This was not the case, but in removing it the chassis was cut up and soon afterwards 'Chitty I' vanished for ever.

After the Count had decided to dispose of 'Chitty II' to leave more space in the Garages at Higham it was thought to have been acquired by Merrick Fowler for the sum of £825. In April 1927 it featured at Glass's

The Benz-engined 'Chitty-Bang-Bang II' when David Scott-Moncrieff (right) was trying to find a purchaser for her. Kent Karslake later sampled the ageing monster for Motor Sport.

The famous advertisement which Scott-Moncrieff, 'Purveyor of Horseless Carriages to the Nobility and Gentry', inserted in Motor Sport *in 1930, the car being 'Chitty II'.*

View from the Bar, somewhere near Weybridge.

Two-Seater Mulliner, 30-98 Special Engine, 185 guineas.

Four-Seater 30-98 Zeiss Lamps in Wizard Condition, £245.

Four-Seater O.H.V. 30-98 F.W.B., Paint Good, £135.

S.A.R.A., 9 h.p. Air Cooled Four Cylinder Practically Unused, £135.

2-litre Supercharged THOMAS Special £195.

45 h.p. RENAULT, would be sold in Plots or Exchange 1½-litre Basinette or anything useful.

You want ALFA ROMEOS. We have them

1930 TRACTA Sunshine Saloon, 295 gns.

90 m.p.h. Monza HISPANO 295 gns.

ASTON MARTIN, 1928 late Ex-demonstration 4-Seater. Just had £100 spent on it, £345.

Scott-Moncrieff

LTD.

The Sports Car Enthusiasts, now of 5, Yarmouth Mews, Brick Street, Piccadilly, W. (Grosvenor 2055).

Used Car Show (and sale) at the Alexander Palace. A playboy of the time, Roddy Williams, had her next and may have bought her at the Show. Then David Scott-Moncrieff, the well-known 'Purveyor of Horseless Carriages to the Nobility and Gentry', and author, took 'Chitty II' on and put her up for sale—one of his advertisements in *Motor Sport* included a photograph of the car published *upside down*, with the caption 'Chitty as seen from the Brooklands' bar'! At this time E.K.H. Karslake had a run in her and wrote of the car for *Motor Sport*, in his 'Veteran Types' articles.

Mr Hollis, a Dover farmer, bought the car in July 1930 and for some time used it on the road. When war broke out he removed it to nearby Sutton from Dover to escape enemy bombs. Here it became derelict and attracted the attention of Peter Harris-Mayes, a wholesale butcher, in 1944. He persuaded Mr Hollis to let him borrow 'Chitty' and had much work done on the old car, including getting Thomson & Taylor's of Brooklands to make a new gearbox for it. When Harris-Mayes collected her from the Track one Sunday morning I rode with him part way on his journey home to Deal after we had been seen off by Ken Taylor himself. After that the new 'owner' ran the old car in a number of events, including the Brighton speed trials, and drove it to Alexander Palace again for a television appearance. However, the repairs had cost him a lot of money, the delicate gearbox was at the mercy of the scroll clutch, so he decided to offer it for sale at Norman Cole's auction at Olympia in July 1969. A bid of £16,500 was made for the car by Mr Harry Resnick of America, but friends alerted Mr Hollis to what had happened. He had parted with 'Chitty II' on a loan basis 25 years earlier and did not wish to incur the expense of legal action. However, Lord Montagu of Beaulieu offered to pay his costs, and for the time being an injunction was obtained to ensure that the Count's old car would remain in this country until the

'Chitty-Bang-Bang II' photographed just after Thomson & Taylor's had overhauled her for Mr Harris Mayes of Deal. The person in the centre of the group looking at the engine is Lt. Col. Clive Gallop, who was Count Zborowski's close friend at the time when these great cars were being built.

The offside of 'Chitty II's' 18.8-litre Benz engine. Note the exposed springs for the 24 overhead valves and the long risers from the two carburettors to the inlet manifolds.

The dashboard of 'Chitty II' as it was in Harris Mayes's time, with the hand-wound starting magneto prominently in view. The steering wheel is an obvious replacement.

hearing. For some eight months it was stored in a crate at Evan Cook's premises in Peckham. It still retained its original registration number, FN5230.

When the case was heard in the High Court, at which I was an ineffectual witness for the plaintiff, a case which had several humorous asides, the judge found for Mr Harris-Mayes and 'Chitty'. It was lost to an American Museum in Cleveland, Ohio. Incidentally after the demise of 'Chitty I' Brian Morgan kindly gave me one of her chain sprockets.

'Chitty II' passed from the American collector to the Western Reserve Historical Society in Cleveland, Ohio, joining the Frederick C. Crawford Auto-Aviation Collection which in the summer of 1992 loaned the famous car to the National Motor Museum at Beaulieu, Hampshire, England. So one of the immortal 'Chittys' returned, if only temporarily, to England.

4. The Martin-Arab

The Humber Company built a team of four cylinder 82×156 mm. (3,295 cc) twin-cam cars for the 1914 Isle of Man Tourist Trophy race, based on the Ernest Henry-designed Peugeots. One of these Humbers won a race at Brooklands, driven by W.G. Tuck, at the fateful August Bank Holiday meeting of 1914.

After the Armistice one of them went with its designer, F.T. Burgess, to the Bentley works at Cricklewood, where it was used as a basis of the new 3-litre Bentley chassis, after which it was put up for sale. All three of the Humbers which had competed in, but retired from, the T.T. were raced at Brooklands after the war, by W.G. Barlow, W.D. Wallbank and Philip Rampon, a wine importer. Only Wallbank's had any success, winning a B.A.R.C. handicap event there in 1929.

Rampon had very little luck with *his* Humber. Its engine blew-up during a 1920 Brooklands race. He is said to have installed another four-cylinder engine, of 102x160 mm. bore and stroke (5,230 cc), but the make of this mysterious substitute has never been established; *The Autocar* referred to it as a Grand Prix Vauxhall power-unit, but the dimensions do not confirm this. Moreover, in a letter he wrote to me in 1947 Rampon denied using any but the Humber engine in his car, until he converted it to take an aero-engine. The mystery is not helped by a photograph he sent me of the Humber at Brooklands with the exhaust pipe on the near side, whereas a distinguishing feature of the 1914 T.T. Humbers was that they had their exhaust pipes on the *off*–or driver's–side. Adding further to the confusion, *The Motor* reported that when, after a lap by the Humber of 95.96 m.p.h., this replacement engine also blew up, it shed a cylinder, but none of the engines likely to have been used had separate cylinders, not even the Sunbeam Arab aero-engine the car had by 1921.

In his letter Rampon did mention two sizes of Sunbeam Arab aero-engines. Is it possible that he managed at first to install the smaller one with the Humber radiator in its usual place, but the exhaust pipe on the nearside?

The fact remains that after these repeated mechanical disasters Philip Rampon decided to follow the aero-engined line and install a V8 240 hp Sunbeam Arab engine in the Humber chassis, renaming the car the Martin-Arab, the work having been done by his mechanic Ernest Martin at Rampon's home garage, which was probably adjacent to his riverside residence at Staines. The bonnet was widened to take this Sunbeam Arab engine and the radiator moved forward. Kenneth Neve, who campaigns the only surviving 1914 T.T. Humber, kindly measured the under-bonnet space in his car, which is 17.75″ wide behind the radiator, widening to

Philip Rampon's Martin-Arab, which had a Sunbeam aero-engine in a 1914 TT Humber chassis, at Brooklands in 1921.

25.75" at the scuttle, the bonnet being 37.5" long. However, in Humber form the engine sub-frame has a width of only 15". The 200 hp Sunbeam Arab aero-engine was 370 mm. wide at the bearer arms, so it seems it could have fitted into the Humber sub-frame.

Be that as it may, after the second serious engine failure at the 1921 Brooklands Easter meeting, the black Martin-Arab with the V8-cylinder 120×160 mm. (14,477cc) Sunbeam Arab aero-engine was entered for the 100 m.p.h. Long Handicap at the Whitsun meeting. It non-started. It also missed the Summer Brooklands meeting but came out for the Surbiton & District M.C. Brooklands meeting at the end of July. It then appeared in August, with Martin as its driver. In the 100 m.p.h. Short Handicap he was heavily handicapped, being on scratch and giving even Hornsted's 21.5-litre Benz and John Duff's 10-litre Fiat a start of four seconds. He was posted a non-runner. Martin then had little success with Rampon's six-cylinder 9.5-litre Berliet-Mercedes, borrowed from the Robinson brothers, and the Martin-Arab was the slowest car in the Senior Sprint Handicap.

Rampon then sold the car to Capt. Duff, taking over and racing his red 11-litre Fiat. Although the identity of the three T.T. Humbers raced after the war has never been established (not for want of trying!), it seems likely that the Martin-Arab was broken up.

5. The Cooper-Clerget

The Cooper-Clerget was undoubtedly intended to follow the role of aero-engined Brooklands racing cars set by Count Zborowski's 'Chitty-Bang-Bang'. It was evolved for Captain Jack Hartshorne Cooper, who with his brother, Major Richard Francis 'Shuggar' Cooper, MC, was a close friend of the Count's. Indeed, perhaps his closest friend at the period in his life when 'Chitty-Bang-Bang' was born.

Capt. Cooper had served in France during the war and had flown with the RAF and had been racing one of the 1908 Grand Prix Mercedes cars at Brooklands (somewhat altered, since it had been built for racing in the Grand Prix) in 1920 and thought to be the ex-Tate car. This 12.9-litre Mercedes, which significantly by then had a Canterbury registration number like 'Chitty', scored a third place at the August Bank Holiday Meeting in the Private Competitors' Handicap race, behind the old 1912 15-litre GP Lorraine Deitrich and the Count in his 1914 4.5-litre GP Mercedes.

At the following Essex M.C. Brooklands Meeting, Jack Hartshorne Cooper's Mercedes won the Essex Lightning Short Handicap race at 90.25 m.p.h., and he came home strongly in third place in the Essex Long Handicap. At the Autumn B.A.R.C. meeting Jack Cooper took two more third places in the ancient Mercedes, in the Lightning Long and Senior Sprint races.

In 1921, the first full season's racing at Brooklands since the war, 'Chitty-Bang-Bang' made her sensational debut and clearly Jack Hartshorne Cooper craved a car of similar power, his old Mercedes having about come to the end of its promising racing career. Indeed, it was a non-starter at the Easter races, although Major Richard Cooper, having caught the racing fever, ran a cream 9-litre Mercedes, presumably a pre-war Ninety model. The close friendship between Count Lou Zborowski and Hartshorne Cooper was seen when the latter, his own car not ready, borrowed Lou's prized 1914 GP Mercedes for the Senior Sprint Handicap and won, even somewhat baulking the Count in 'Chitty' in the process. It seems obvious that, not to be outdone, Jack Cooper was having his chain-drive 1908 GP Mercedes converted into an aero-engined car, to which its short wheel-base may not have been altogether suited.

It seems almost certain that the work was done somewhere in Canterbury but it is thought not at Zborowski's Higham workshops; the new engine either one of his or obtained from one of the aforesaid war-surplus dumps. It was, however, a rather unusual choice, a rare V8 Clerget of 140×160 mm. (19,704 cc), reputed to produce 200 hp. Clerget were better known for their rotary aero-engines, which W.O. Bentley

As far as can be ascertained, the Cooper-Clerget was never photographed, perhaps because Capt. Hartshorne Cooper was killed in it before it was ever raced. But it seems likely that the Clerget aero-engine was installed in this 1908 GP Mercedes which Capt. Cooper had driven previously.

improved into his BR1 and BR2 rotaries, working at Humber's in Coventry. The vee-formation Clerget engine was probably better suited to the Mercedes than an in-line six.

The Mercedes, now called the Cooper-Clerget, was entered for the 1921 Brooklands Whitsun meeting, in the Lightning Short and Long Handicap races, the Senior Sprint Handicap and the Sweepstakes. The black car had been put on the same mark as 'Chitty-Bang-Bang', in spite of its rather lesser engine capacity, a handicap speed of 110 m.p.h. Apparently the work was not completed until early in May (racing was on 16 May) and after a few test runs on the road, a lorry, perhaps Zborowski's solid-tyred Mercedes, took it to Brooklands on the Tuesday previous to the Whitsun Bank Holiday meeting. On test the next day it bucked about seriously. The back springs were altered, new dampers fitted, and runs in the afternoon showed an improvement. This encouraged Jack Cooper, with Harold Easton as his passenger, to open-up. Coming down to the Fork the Cooper-Clerget appeared to leap off the ground (had something broken and dug-in?). It then skidded, spun round, crashed into the iron railings bordering the track, and overturned. Hartshorne Cooper was pinned beneath the car, to be pulled out by Zborowski, who was the first to reach the scene of the calamity. He was taken to Weybridge Cottage Hospital but died from a fractured skull half-an-hour after admission. Easton escaped with minor injuries.

At the inquest next day a verdict of 'Death by Misadventure' was returned. Although the family seat was in Derbyshire Jack Hartshorne Cooper is buried in Weybridge Cemetery. Dick Cooper, who had intended racing Capt. Duff's big Fiat 'Mephistopheles', never did so and he never spoke of his brother's accident. An aero-engined car had taken its toll.

6. The Wolseley Viper

Alastair Miller became interested in motoring before he came of age, owning a T.T. Rudge and other motorcyles. He went into marine insurance but left to enter the motor trade. When war broke out he joined the Irish Guards and the RFC, flying BE2c, Avro and DH6 aeroplanes, with the inevitable 'prangs'. He was soon invalided out and put on light home duties (his elder brother, a Lieutenant in the Grenadier Guards, was killed in action at Ypres in 1914). After the Armistice Alastair Miller formed his own motor business, trading in used cars and rebuilding ex-RAF Crossley tenders, etc. He acquired the two 1914 GP Opels which had remained in this country throughout the war, with one of which Sir Henry Segrave started along the path to fame. Miller was associated with racing these cars during the 1920 Brooklands season, and in 1921 undertook to run a Competitions Department for A.J. McCormack of Wolseley Motors Ltd. For this purpose he took over some sheds on the Byfleet side of Brooklands track. The first car he devised for his Wolseley fleet consisted of the very fully streamlined little 'Moth' single-seater, which was based on the production o.h.c. Wolseley Ten light car. This was ready in time to run at the last B.A.R.C. Meeting of 1921. Later another, virtually identical, 'Moth' was built and a larger Wolseley two-seater was added to the fleet in 1922, in which either a 2-litre or a 2.7-litre engine could be installed, as required (a similar car being sent out to the Argentine), while for the second J.C.C. 200-Mile race in 1922 Miller prepared a two-seater Wolseley Ten. In normal guise these Wolseleys were very sluggish little cars—one authority mentions a best cruising speed of about 26 m.p.h.—so it is greatly to Miller's credit that he eventually had a 'Moth' lapping the Track at over 88 m.p.h. and still winning as late as 1930. These various Wolseleys between them gained an impressive number of successes and broke innumerable records—someone writing about Brooklands in *Motor Sport* has remarked that as different periods of the earth's history are known as 'ice', 'heat', etc., so this era of Brooklands might have been known as the 'Wolseley-age'. Even when nothing much was happening, there was nearly always one of Miller's Wolseleys going round and round.

The Wolseley Viper was rather different. Ever since the Brooklands Track authorities had instituted the 'Lightning' Handicap races in 1914, the biggest and fastest cars attracted the most attention and caught the full imagination of Brooklands *habitués*. Giant pre-war Mercedes, Fiat, Benz and GP Lorraine-Dietrich racers were enlivening the Bank-holiday meetings and other aero-engined monsters, like the Cooper-Clerget and the Martin-Arab were under construction, while Zborowski was building a second 'Chitty'. Miller felt he could not rest content until he, too, had a really big and exciting racing car.

He was helped in this ambition by the fact that, during the war, the Wolseley Company had manufactured Hispano-Suiza aero-engines under licence, these vee-eight engines with a shaft-driven camshaft over each bank of cylinders being known as Wolseley Vipers. All that Miller needed was a chassis able to accommodate one of these formidable power units. Whereas other builders of giant Brooklands cars were obliged to haunt war-surplus stores in the hope of picking up suitable inexpensive aeroplane engines, Miller no doubt found brand-new Viper engines in the Birmingham stores of Wolseley Motors Ltd.

There was a splendid story, to the effect that at a dinner party at which HRH the Prince of Wales was present, Miller referred to his craving for a giant racing car. The story continued to the effect that the Prince recalled a big pre-1914 Napier brake at Sandringham, formerly used for Royal shooting parties, and said he would arrange for Miller to have it. This duly came about, but when the old brake, converted into the Wolseley Viper, was careering round Brooklands, so the story had it, King George V suddenly demanded to know what had become of his old Napier and the Prince was horrified that he might find out! However, I recently took the opportunity of asking the late Duke of Windsor's secretary if he could confirm this and was told that nothing is remembered of it and that the King had never used a Napier anyway.

Nevertheless, Alastair Miller definitely used a Napier chassis for his

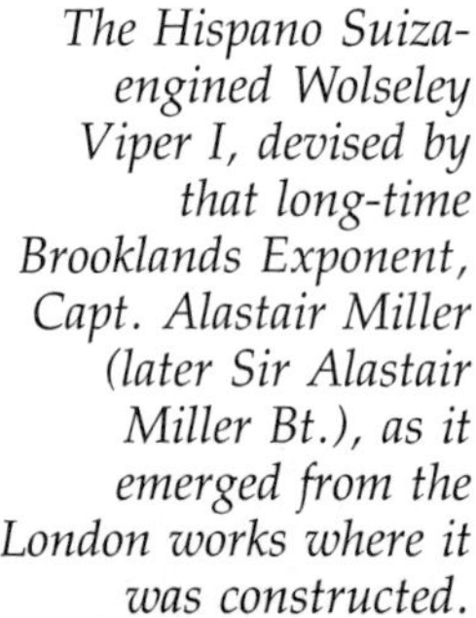

The Hispano Suiza-engined Wolseley Viper I, devised by that long-time Brooklands Exponent, Capt. Alastair Miller (later Sir Alastair Miller Bt.), as it emerged from the London works where it was constructed.

aero-engined racing car. He would probably have preferred to use one of Wolseley manufacture, but perhaps the cantilever rear springing and worm-drive back axle of their biggest model, the 20 hp, put him off, or perhaps its construction was too flimsy (or, with post-Armistice production hold-ups, Wolseleys may not have been prepared to produce one); presumably a pre-war 50 hp Wolseley chassis wasn't forthcoming, either. So a Napier it had to be, Miller having no scruples about entering the car as a Wolseley, even if it consisted of an Hispano-Suiza engine in a Napier frame.

The Napier Company has been unable to tell me anything about this chassis (No. 2482), but it was, I think, a 60 hp L-type. It stood high, on ½-elliptic springs above the axles, with the steering tie-rod in front of the deeply-dished front axle in typically Napier style, and as Miller made no attempt to lower it, the Viper was one of the loftiest of the Brooklands cars. It was built for Miller by the Line brothers in a workshop in London. The engine was a Type 34 200 hp 120×130 mm. 90 deg. vee-eight Hispano-Suiza aero-engine, with a flange over its propeller boss to accommodate a clutch. This was the engine which powered the RAE-designed SE5 fighters, and Miller's was probably a direct-drive high-compression version which developed just under 200 b.h.p. and weighed 445 lb., the aluminium cylinders and o.h.c. valve gear being typical of its creator, Marc Birkigt. (It seems almost certain that the geared version of these engines, also rated at 200 hp, would not have been used in the car, as the drive offtake would have been too high and

Alastair Miller in the Brooklands Paddock in his unwieldy but effective hybrid racing car.

the direction of rotation incorrect.) New, these engines cost over £1,000 each, but I do not suppose Miller paid that for his! A shapely two-seater body with wind-deflecting cowls on the scuttle, a full-length undershield, and a rakish nose cowl over the radiator was made for the car by a radiator company in Munsden Road, Hammersmith. Apparently little was done to the ancient chassis, apart from putting in a new E.N.V. crown-wheel and pinion of KE805 steel, no doubt of a suitably high ratio, making new gears for the gearbox, which now had only two forward speeds, and presumably converting the hubs to take centre-lock wire wheels. Aero-engines do not have clutches, so Miller had to find one. At first a clutch from a Crossley tender was tried but this proved unsatisfactory. Count Zborowski, then the foremost amateur-builder of aero-engined racing cars, was appealed to and he came up with a Hele-Shaw clutch, which solved the problem. The swept-volume of the Viper was 11,762 cc, small by the standards of other aero-engined cars but half as large again as the power units of the bigger production models of the period, so this imposing car, which weighed nearly 33 cwt., was rightly regarded as an exciting creation. It was clearly to be more Miller's personal venture than the other Wolseleys, which had their bodies made in Birmingham and were virtually 'works' cars.

The Wolseley Viper was ready for the 1921 Autumn Brooklands meeting, along with the first Wolseley 'Moth'. Miller brought it out, in its iron grey finish, for its first race, the Second 90 m.p.h. Short Handicap. The handicapper was not taking any chances; the Viper was on scratch with Segrave's 3-litre straight-eight Sunbeam. But the car was not running

Miller going for standing-start class records with the Viper, and getting a push from a helper.

properly and it faded away after doing its standing lap at a mere 66.38 m.p.h. It was not a happy day for 28-year-old Miller, because in his next race a Bugatti swerved down the banking into his 'Moth' and overturned. The Wolseley was quite severely damaged, but by borrowing a front axle from a standard Wolseley Ten, Miller was able to bring the little car out for its next race, the 75 m.p.h. Long Handicap, from which, however, it retired. Nor had he any better success with the Viper in the '90 Long', for, again on scratch with Segrave's Sunbeam, he retired after lapping at 78.16 and 86.62 m.p.h.

Although the Viper had not made much impression on its first racing appearance, at least this improved its chances on handicap for the 1922 season. Early in February Miller had the car out for the edification of some motoring journalists. It was described as having a 240 hp Hispano-Suiza engine; the Press usually exaggerated the power of the big racing cars, but it is possible that the engine had been souped up to some extent, although it ran on ordinary petrol. Now painted green, the Viper was tow-started by a big touring car 'invisible when sitting in the pilot's seat, owing to the length and height of the monster,' the frosty stillness of the Track being 'awakened by a shattering roar as the eight cylinders leapt into life,' exhausting from long open pipes at each side of the car. Miller did some fast laps, the Viper diving alarmingly down the bankings, due to the weight being well forward, which today we would call understeer, after it had been steered away from the top of the track. Both Miller and Don testified to the Viper's habit of wandering up and down the bankings, but Miller claimed it was one of the safest cars on the

The vee-eight Wolseley Viper carries the appropriate emblem on its radiator cowl!

Track, for anyone who knew how to drive it.

As the car was domiciled at Brooklands, Miller had no problems about transporting his big racer to and from meetings, but he made frequent journeys down from town in Wolseley Ten and Fifteen 'slave' cars (he had not yet gone to live in the bungalow which later became Parry Thomas's permanent residence) to work on it, helped by Col. Stewart, Jack Woodhouse, who was riding Miller's Martin in motorcycle races, and the Line brothers.

Later that month the car was sent up to London to have shock-absorbers fitted, and in March, while practising, a valve broke and destroyed a piston and cylinder due to the car going out with no oil in the dry-sump oil tank. After a great deal of work this was rectified; I think this may have been when the Wolseley Viper W4A wet-sump engine was installed. These engines were rated at a nominal 150 hp at 1,500 r.p.m. for a weight of approximately 440 lb. but as the bore and stroke were the same as those of the 200 hp Hispano-Suiza, output in the car would be identical. Wolseley charged £814 for these engines when new, but I imagine Miller got one for nothing. The resuscitated Viper was entered for the 1922 B.A.R.C. Easter meeting. Now painted blue, it was too heavily handicapped to get a place; the entrants were Miller and a Capt. R. Wilson, and its best lap was at 98.23 m.p.h. Incidentally, on the road at this time Miller often used a closed Wolseley 15 and his father had a new Wolseley 20.

For the Whitsun races the Viper was again repainted—black. In the 100 m.p.h. Short Handicap, Miller started from the 22 sec. mark and was caught by Segrave, who, in the scratch 5-litre Sunbeam, found the Viper so high up the banking that he had to wait until they got to the Finishing

Miller and his passenger, Capt. Oliver Stewart, the aviation writer, look back at the opposition as they approach the Fork during a Brooklands race.

The well-known racing driver Kaye Don, who also raced the Wolseley Viper, with it in the Paddock at Brooklands.

straight to pass, the Viper finishing in second place, after lapping at 88.10 and 100.61 m.p.h. In the '100 Long' Segrave was re-handicapped, Miller had 33 sec. start and the Viper, now lapping at 94.06, 107.34 and 106.42 m.p.h., won by about 1.5 miles, at 102.03 m.p.h. Two days later the Viper broke the Class H f.s. kilo record at 118.29 m.p.h. but finished its run with only one tyre tread intact. The tyres just wouldn't last long enough for it to take the mile record.

The car took part in one race at the Duke of York's meeting, again running high on the bankings, without success (it was 4th) but at the next Essex M.C. meeting Miller won the 'Lightning Short' at 99 m.p.h., and Kaye Don was second to Thomas's Leyland in the 'Long', having overtaken Thomas before losing a tyre tread; the Viper apparently achieving its fastest race lap ever, at 112.68 m.p.h. For the August meeting Kaye Don again drove the Viper, the nose and tail of which were now picked out in red. He drove with great skill from the scratch mark in the 100 m.p.h. Short Handicap, winning at 101 m.p.h., after lapping at 92.59, and 111.42 m.p.h.; a fatal accident caused racing to be abandoned before the car's next engagement. Late in September L.C.G.M. le Champion, who later raced cars as versatile as an Akela G.N. and the Isotta-Maybach, brought the Viper out for an attack on long-distance class records but was soon obliged to give up, due to anxiety about the oil-pressure and carburation trouble. This is probably why the car wasn't entered for the Autumn meeting. But it reappeared at the Armistice meeting, causing some consternation by catching fire; but getting off the line well in the Lightning race, in which, however, it was unplaced. In its first full season the Viper had proved itself an effective two-mile-a-minute car, while the danger of driving these huge racers had been emphasized by the accident to 'Chitty-Bang-Bang' and by Duff going over the top of the Members' banking in the big Benz.

The Wolseley Viper also made f.t.d. at the 1922 Westcliff speed trials, in 34.8 sec., driven by Capt. Miller, and also won a Match Race against the fastest motorcyle. At the 1922 Southsea Speed-trials that year it ran unsuccessfully, driven by Miller. It was also demonstrated at that year's Eastbourne speed trials.

By 1923 Miller was fully occupied with his Wolseleys, although George Newman and Woolf Barnato had taken over some of the driving. Incidentally, the 'Moths', named after some chorus-girls with whom Miller had made friends, were numbered 'Moth' I and II, the bigger Wolseley I with the 2-litre engine, III in 2.7-litre form (II was presumably the car sold to an Argentinian enthusiast for racing in that country), the 200-Mile Ten and the Viper both being numbered I, probably in case any more of their kind were built. Apart from loaning his Wolseleys to other drivers, I suspect for a fee, Miller had persuaded Percy Cory, the actor, that he should have a share in racing the Viper. He entered it for the 1923 Easter Brooklands meeting, with Kaye Don as driver, and went as passenger, a fine way of enjoying a Bank Holiday, which was open to owners of two-seater racing cars of this era. Alas, in this case the outing was hardly enjoyable! The Viper, giving four seconds to Thomas in Howey's Leyland in the Founders' Gold Cup race, lapped at 87.27 and 108.98

Kaye Don makes a mistake!—Finding that the brakes of the Viper would not stop him from a trip over the top of the Members' banking, he finished his race by rubbing the car along the bank below the Test Hill, at the top of the Finishing Straight. Cory, who had entered the car, hit his nose on the wind-deflector in front of him and the oil pump was torn off. Otherwise, not much blood was spilled.

m.p.h., leaving Thomas behind; it was rapidly catching Cook's T.T. Vauxhall when Don discovered that he was going too fast to pull up after the finish. Instead of trying to take the turn on to the banking, as Duff and others had done with unhappy results, he decided to scrape along the sandy wall of the embankment on the right of the finishing straight. The Viper pulled up safely but poor 'Pop' Cory bumped his nose on the scuttle wind-cowl. Miller says Don knocked the gear lever into neutral when reaching out for the handbrake and that he should have switched off; Don says a pin broke, rendering the Viper's small-diameter rear-wheel brakes inoperative. The car was scratched from its next race because the water-pump had apparently been torn off. It was absent from B.A.R.C. meetings for the rest of the season.

For 1924 Cory, perhaps wisely, drove one of the small Wolseleys. D. Fitzgerald entered the Viper, now painted red (with a viper rampant, if that is the term, in a circle on each side of its long radiator cowl) and with cylindrical silencers incorporated in its exhaust pipes, for Norris to drive. This was a successful combination, for at the Easter B.A.R.C. meeting second place was secured in the Lightning Long Handicap behind Campbell's 5-litre Sunbeam, the Viper lapping at 109.94 and 107.34 m.p.h. but losing a tyre tread, which dislodged the offside exhaust pipe. At Whitsun, Norris was unplaced in the Gold Vase race, the best lap

Viper I being brought back to the Paddock after Kaye Don's contretemps at the Easter 1923 Brooklands' races.

speed down to 102.69 m.p.h., and he retired from the 100 m.p.h. Long Handicap. Between these appearances, Norris had used the car to take the Class H 250 kilo and 3-hour records, at 85.37 k.p.h. and 53.08 m.p.h., respectively, there being useful bonus payments for any records which could be raised, as Miller well knew. The Viper weighed-out at nearly 35 cwt. on this occasion. It was aiming for the 6-hour record, but something was obviously amiss, and after three hours it had covered only just over 161 miles. It was unplaced at a Club meeting, again losing a tyre tread, and thereafter Norris concentrated on sharing the driving of the two Bianchis, very profitably, with Miller.

By 1925 Kaye Don had taken over the running of the Viper. Unplaced in the 'Lightning Short' at the Summer Brooklands meeting, although lapping at 91.98 and 111.17 m.p.h., he non-started in this race in August but came second in the equivalent event at the Autumn meeting, although the lap speeds were down to 87.07 and 110.68 m.p.h. Perhaps, though, these intermittent appearances had aided the handicap, for the Viper was on limit in this race, Cobb's winning 10-litre Fiat giving it six seconds start.

The following year, the Viper, although composed of pre-war and wartime components, continued to serve K. Don well; incidentally, in 1926 people were still entering the aero-engined monsters, and even building 'new' ones. At the Easter Brooklands meeting the Viper's lap-speeds were held down to 78.79 and 96.15 m.p.h., but at Whitsun, in the Gold Vase race, Don took it round at 93.97 and 110.43 m.p.h., finishing second, a length behind Miller's 5-litre Sunbeam, its old-time rival, both cars starting together, the Sunbeam accelerating more slowly but being quicker thereafter. The Sunbeam failed to start in the 100 m.p.h. Long Handicap, and although Don was re-handicapped, the Viper came through from the 3-sec. mark to win by quarter of a mile at 104.91 m.p.h., after doing its s.s. lap at 93.97 m.p.h. and both flying laps at 111.92 m.p.h. (For this he received a cup worth £30, having paid £5 to enter.) Either because the old car, described by *The Autocar* as 'vicious but venerable', was getting a bit long in the tooth or to conserve his handicap, Don entered for only one race at the Summer B.A.R.C. meeting. He came home second to Howey's Ballot in the 'Lightning Long', lapping at 94.68, 111.42 and 110.43 m.p.h. Adopting the same tactics for August, Don made a bad start (s.s. lap at 67.21 m.p.h.) so that subsequent laps at 110.43 and 109.94 m.p.h. availed him nothing. A similar appearance at the 1926 Autumn Meeting produced laps at 92.51 and 106.65, after which the Viper fizzled out.

That concluded the Viper's racing career but it remained active at Brooklands, being used by Avon for tyre testing. There is also a story to the effect that it lent its engine to a racing aeroplane on one occasion, this being removed very expeditiously after the flying race had ended at Brooklands and being re-installed in the Viper in time for it to race that day, but I have been unable to tie up dates on which this could have happened.

After 1930 the authorities banned the older cars from public appearances on the Track. In 1931 the Viper was offered to me, with sound tyres, for £25. In those days £25 was impossible to raise and I had reluctantly to forgo the pleasure I would have derived from keeping the old warrior in one of the inexpensive lock-ups at the Track and taking it out for joy-rides on fine evenings, as other folk went up in their aeroplanes. In the same way the advent of another World War forestalled my ambition to live in the stone house on the aerodrome side of Brooklands, which had been unoccupied for some years.

What became of the Wolseley Viper? This I never discovered. One can only suppose that it was a victim of the wartime scrap drive. But it seems odd that such a memorable car did not cause some comment as it went to its final resting place.

7. The Isotta-Maybach

This car started life as an Isotta-Fraschini. In 1907 Nazzaro won the Grand Prix on a Fiat, and Italian cars generally were doing extremely well in motor racing. Minoia's Isotta-Fraschini won the Coppa Florio at 64.7 m.p.h. from a couple of Benz cars, and the following year Trucco won the Targa Florio, averaging 37.24 m.p.h. for 277 miles of the notoriously difficult mountain course, while these overhead camshaft cars took important honours in America, including second place in the Vanderbilt Cup race. Different rules governed these races but for the 1907 Kaiserpreis race Isotta-Fraschini built three chain-drive 8-litre cars with a bore and stroke of 145.4×120 mm. which, however, were not particularly successful, although, as stated, they made up for this in the Italian races.

As so often happened, some of these road-racing cars found their way to Brooklands. In 1913 an Isotta-Fraschini which had a rather bigger engine than the Kaiserpreis cars was singularly unsuccessful there, but in 1914 Humphrey Cook, who much later financed Raymond Mays' E.R.A. project, had a good season with a standard 100 hp chain-driven model endowed with racing bodywork. Then, after the Armistice, the late Ernest Eldridge, who was an avid big-car enthusiast until he brought a 2-litre Miller back from America and built the 1.5-litre Anzani-engined Eldridge Specials, got hold of one of the Kaiserpreis cars and entered it for the 1921 Brooklands Summer meeting. Painted in grey, it lapped at 89.9 m.p.h. on its first public outing. This was not fast enough to secure it a place, and although Eldridge lapped almost as fast in his next race, a 75 m.p.h. Short Handicap, in which he found himself on scratch, giving a minute's start to George Duller's Silver Hawk, doing 77.52 and 89.41 m.p.h., he was again unplaced. He non-started in the '100 Long', as reserve entries were not called, but was out again, still on scratch, for the '75 Long' and, getting the 14-year-old Isotta-Fraschini round at 77.39, 90.72 and 88.15 m.p.h., finished second behind T.B. Andre's Marlborough.

This was an excellent come back for the car, the engine size of which was now declared at 146×120 mm. (8,036 cc), so it may have been rebored; water cooling had been contrived round the valve ports. No doubt pleased by this debut, Eldridge was out again at the 1921 August Bank Holiday Brooklands meeting. He started with a good handicap in the 100 m.p.h. Short Handicap, the only car to leave before him being a 40/50 Rolls-Royce. The Isotta lapped at 77.9 and 87.38 m.p.h., travelling so high on the bankings that it baulked faster cars, but could not retain its lead. These comparatively low speeds preserved its chances on handicap but in the '90 Short' it dropped out after a standing-lap at a mere 69.31 m.p.h. It was revived in time to contest the '100 Long', in which it lapped at 89.74 and 90.88 m.p.h. without getting a place. However, at this meeting, at which Eldridge had entered for four events, he got his reward in the 90 m.p.h. Long Handicap, in which the Isotta-Fraschini started from the 48 sec. mark, lapped at 76.01, 88.94 and 91.38 m.p.h., and finished third behind Bedford's famous Hillman and Cook's 30/98 Vauxhall.

Thus encouraged, Eldridge ran the car in the Autumn Brooklands races, its speed higher than ever before, for in the '90 Short' he got round at 77.65 and 91.72 m.p.h. This was not quick enough to get an award, however, and although Eldridge lapped at 70.02, 93.44 and 95.78 m.p.h. in the '90 Long' he was still unplaced.

It was probably this comparatively unsuccessful season that prompted Eldridge to re-form his old Isotta-Fraschini. He was intrigued by really big cars, and is thought to have had a hand in the building of the 21-litre Maybach-engined Metallurgique owned later by Douglas Fitzpatrick. So perhaps it was no surprise to Eldridge's friends when he got to work in the Vauxhall Bridge Road, lengthening the chassis of his 1907 Isotta so that he could install therein a 20.5-litre six-cylinder Maybach airship engine. The work was carried out by Aeromotors Ltd, a firm formed *circa* 1921 for the purpose of supplying engines for marine, stationary, lorry and automotive purposes. The company also offered to undertake marine installations if engines were supplied, and had for sale war-surplus aero-engines of 20 to 200 hp, such as Djinn, Wolseley, Viper, Hispano, B.M.W., Buick, Green, Maybach, Sunbeam and Amazon, costing upwards of £75 each. Confusion exists as to how greatly the Maybach engine installed in the Isotta was modified, and quite what type it was, and whether, in fact, Eldridge owned Aeromotors or had separate premises elsewhere in the Vauxhall Bridge Road.

What is indisputable is that a big Maybach engine was installed in the Isotta. It was a normal 1910/11 Type AZ of 160×170 mm. (20,500 cc) with four valves per cylinder in T-head formation. Two Claudel Hobson carburettors replaced the carburettors fitted at each end of the inlet manifold on the original engine, the pistons of which were rebalanced and fitted with floating gudgeon-pins, and tie-plates inserted between the separate cylinders in an endeavour to obviate leaking water joints; possibly the water inlet and outlet feed to the cylinders were rearranged. These engines developed 180 b.h.p. at 1,200 r.p.m. in standard form. There is no doubt about Eldridge having taken the right decision. Apart from the fact that we do not know how much more useful life remained in the 15-year-old racing engine which had been removed from the car by 1922, it was essential to lap at well in excess of 100 m.p.h. to obtain the maximum of excitement and a decent chance of success in the 'Lightning' races at Brooklands. This the pre-war aviation motor in its pre-war chassis—the chain-final drive of which, with bigger sprockets, was well suited to the new power plant—enabled Eldridge to do. Indeed, eventually the Isotta lapped faster than Count Zborowski's 'Chitty-Bang-Bang I' with its overhead valve Maybach engine of some 2.5-litre greater capacity.

Aided by his mechanic Jim Ames, Eldridge completed the conversion, after which the car was taken to Jarvis of Wimbledon to be fitted with a new black two-seater short-tailed racing body. Shields protruding from the offside of the bonnet enabled the carburetter intakes to remain in still air, as in modern theory; the dumbirons were exposed, brake and gear levers were well raked backwards, outside the cockpit, but the occupants were permitted the luxury of aero-screens. The car was entered for the Brooklands Easter meeting of 1922 as an Isotta-Fraschini.

Its first appearance was in the Lightning Long Handicap. Eldridge perhaps finding that he needed some distance to work up to full speed—was it not Kent Karslake who observed that old engines were not much good until they were warm, but lost power if they got too hot, so were at their best when they were just nicely warm?—for he had not entered for the 'Lightning Short'. Of the five cars billed to contest this race the 350 hp V12 Sunbeam and the 1912 Lorraine-Dietrich shed tyre treads, and Zborowski's 5-litre Ballot, of which much was expected, non-started. So in its first race the Isotta-Maybach, lapping at 82.53, 101.43, 101.02 and 100.82 m.p.h., won at 95.25 m.p.h., from Cook's fast Vauxhall, which it caught on the final lap. An hour and a half after this satisfactory debut the car was out again, for the '100 Long'. But it had already earned a stiff handicap, so Eldridge contented himself with lapping at 87.94, 92.94 and 92.59 m.p.h. and was out of the picture. Incidentally, the Isotta's 'new' engine was started by turning it over with a crow-bar inserted into a kind

Ernest Eldridge in the enormous Isotta-Maybach in the Brooklands Paddock.

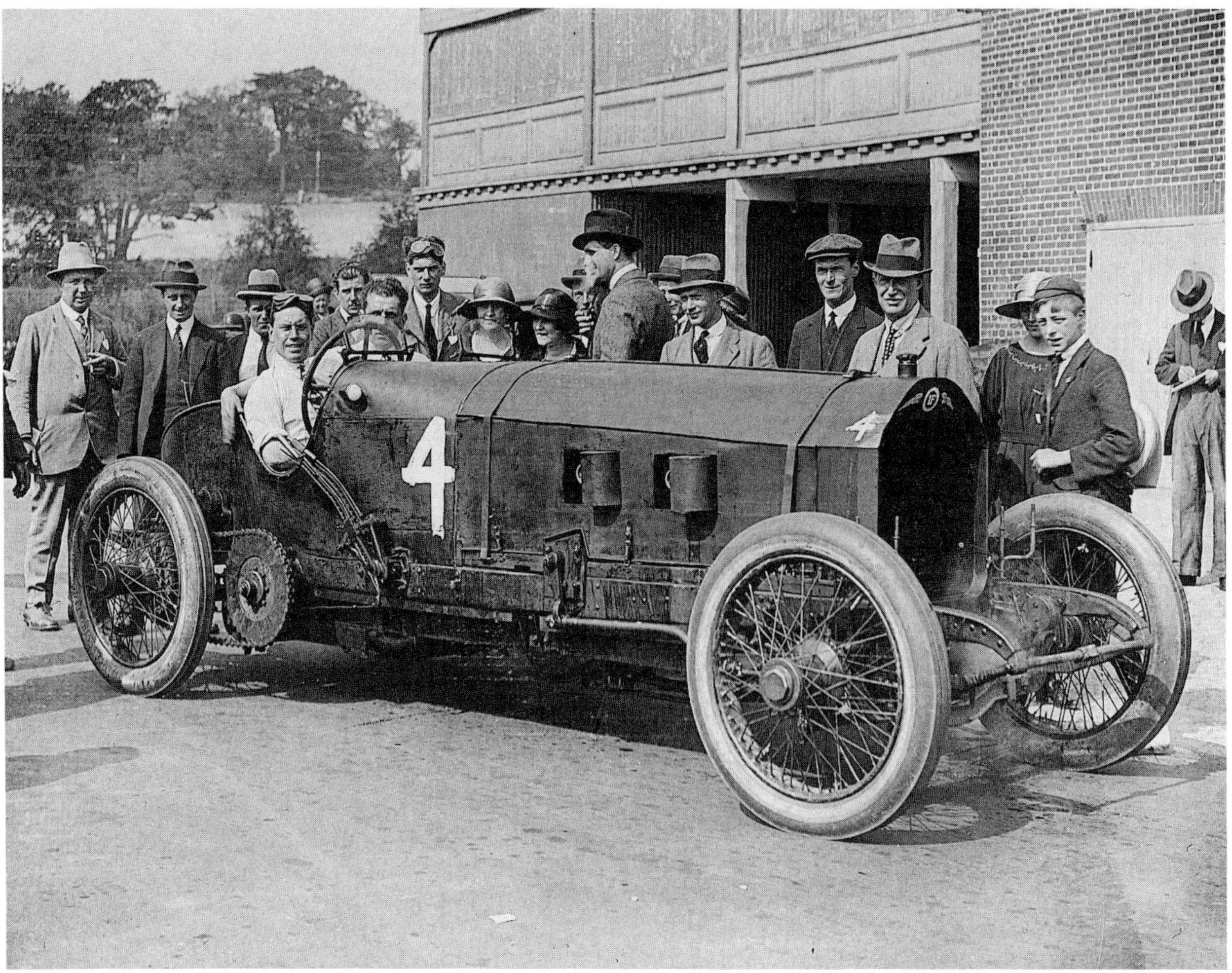

of box-spanner or sleeve which engaged a pin on the nose of the crankshaft and was supported at the front by the dumbirons cross-bar, which, even in 1922, was described as 'quite reminiscent of the old days'. On its first appearance *The Motor* had called the revised car an 'Isotta-Benz-Maybach, or whatever its full title may be', but for the Brooklands May meeting Eldridge still continued to register it as an Isotta-Fraschini. This time it was entered for the Lightning Short Handicap, and although it was going faster than before, lapping at 90.63 and 104.19 m.p.h., it was unplaced. It then failed to get to the start of the '100 Long' but made up for this by winning the Lightning Long Handicap at 102.67 m.p.h., from Zborowski's Ballot, which entailed lapping at 90.63, 109.7, 108.03 and 105.29 m.p.h. Incidentally, crash helmets were not much in evidence in those days, and both occupants of the Isotta were bare-headed.

At the Essex M.C. Royal meeting at the Track—at which, incidentally, some £3,500 was raised for charity—the Isotta faded out, owing to gear trouble or an obstruction in a fuel pipe, according to which report you favoured. But at the Brooklands Whitsun meeting of 1922 the old car was well on form, lapping at 95.21 and 107.1 m.p.h. in the 'Lightning Short'. This failed to get it into the results but in the 'Lightning Long' Eldridge came home first to an easy victory over Parry Thomas's Leyland Eight, averaging 98.75 m.p.h., the lap speeds being 92.06, 105.97 and 100.61 m.p.h.

After this success Eldridge drove the 10-litre Fiat in club races, perhaps while the big Isotta was being improved with a full-length undershield, a longer airship-shape tail and air outlets at the top front of the bonnet. The radiator, however, was still uncowled. Up to this time the Isotta had been easy on tyres but at the August Bank Holiday Brooklands meeting it threw a tread, so missed the 'Lightning Short' and then, in the 'Long' had some difficulty in maintaining second place behind Brocklebank's pre-war Peugeot, having been on fire for much of the race, despite which it lapped at 94.82, 108.27 and 103.54 m.p.h. The car's next competition appearance was at the Southsea Speed Trials at the end of August 1922, an event under the patronage of HRH The Duke of York. Eldridge fitted headlamps for the drive down, which remained on the car during its runs. It was second to Zborowski's 300 h.p 'Chitty-Bang-Bang' in the unlimited class, covering the kilometre in 32.2 sec. (69.47 m.p.h.) to 'Chitty's' 30.6 sec. Eldridge also ran at the 1922 Eastbourne & D.M.C.C.'s speed-trials, where demonstration runs by Brooklands' cars were invited. Here, over an unfettered half-mile of the King Edward Parade, the Isotta-Maybach achieved a time of 22.4 seconds, equal to an average speed of 88.22 m.p.h.

Next, at a minor A.C.U. meeting at the Track the Isotta retired from its first race but reappeared to win its heat in a one-lap Sprint and, in the final, just failed to catch the winning Morgan 3-wheeler, in spite of flinging a nearside tyre tread, which did not seem to make any difference to the car's stability. This business of the car having tyre trouble is rather interesting, because after the B.A.R.C. August meeting Eldridge had told Continental that he was very satisfied with their tyres, on which he had covered four laps at 110 m.p.h., finding the covers hardly worn, while they seemed to have cured any tendency for the car to roll. The speeds seem to be considerably exaggerated unless practice laps were done at higher speeds than were used in races, while before these remarks were written tread-throwing was being experienced. The tyres used were 880×135 Continentals, changed later for 35×5 straight-side Continentals.

The car definitely had its revised bodywork in time for the Brooklands Championship meeting at the end of 1922. In the class-heat it was narrowly beaten by Thomas's Leyland; in the handicap races which followed Eldridge was second to Park's T.T. Vauxhall in the Essex 'Lightning Short' but faded out when in second place in the 'Long'.

There remained only the Autumn and Armistice Brooklands races before the 1922 racing season came to an end. The Isotta-Fraschini got from Pond to Fork at 92.76 m.p.h. in the 17th Lightning Short Handicap but lost a tyre tread and retired. It had started from the same mark as Zborowski's Ballot and the Leyland and had given only three seconds to Barlow's 21.5-litre Benz which went over the banking later in the day. It was scratched from its other autumn races. At the Remembrance Day meeting it started slowly and was no match for its former rival, Thomas's Leyland. However, in September, at a demonstration run at the Eastbourne Speed Trials, Eldridge had covered the half-mile in 20.4 sec (88.22 m.p.h.), beating the Leyland by 3.4 sec.

After this Eldridge gave up driving the Isotta-Maybach and passed it on to L.C.G.M. Le Champion, who had been driving an ancient 4.5-litre Schneider during the 1922 season. Le Champion, whose real name was Le Masurier, was a great enthusiast, frequently passengering in 'Chitty'; he later became interested in spiritualism, writing a book of short stories based on his beliefs. His mother had a place at Crick, near Rugby, and the appearance of her son driving his vast racing cars (in due course he took over 'Mephistopheles' from Eldridge) on the roads near there became a local legend. Eldridge only abandoned the Isotta because he

L.C.G.M. Le Champion, a later exponent of the Isotta-Maybach, poses with it in a photograph in which the under-bonnet machinery appertaining to its side-valve 20.5 litre Maybach engine can be studied.

had discovered an even more formidable car, the famous Fiat 'Mephistopheles'.

So the Isotta-Maybach came into Le Champion's possession. He painted it red and the bonnet was brought to a blunt point behind the radiator to isolate the latter from the engine compartment, with a duct on the off side taking warm air to the intakes of the carburettors, which were now within the bonnet, an extreme form of fairing used by Thomas on the Leyland Eight and for some of the racing light cars; this modification may have been made the previous season. Rubber grips were fitted to gear lever and brake lever and a Motometer thermometer to the radiator cap. Electric sidelamps were retained on the scuttle, so that the car could be driven on the road after darkness had fallen.

Following some trials at Brooklands early in March 1923, when he nearly came to grief because the steering connections had not been pinned, but continued to lap at 80 to 90 m.p.h. and make many ascents of the Test Hill, the new owner's first appearance with the car was at the Essex M.C. Kop hill-climb later that month. Kop was an almost straight road, so it was possible to run these dubiously-braked monsters there. Le Champion made second-fastest time of the day in 30.0 sec., being beaten by Cook's T.T. Vauxhall, which was one-fifth of a second faster. The big car looked deceptively steady, but in fact the front wheels were skidding slightly.

Next it was back to Brooklands for the Easter races. Starting from scratch in the 100 m.p.h. Short Handicap the car, which continued to race as an Isotta-Fraschini, lapped at 86.32 and 111.17 m.p.h. to no avail. Then, again on scratch in the equivalent long handicap, in company with Howey's Leyland Eight, the Isotta, lapping at 90.99, 104.19 and 109.22 m.p.h., caught Joyce's very fast 1.5-litre A.C. and Major Ropner's 30/98 Vauxhall 'Silver Arrow', to win at 100.93 m.p.h. Le Champion was getting good value from his new toy!

The car was driven unsuccessfully in a race at the Ealing & District M.C.C. Brooklands meeting, but at a Surbiton M.C. Brooklands meeting it won the Surrey Lightning Short Handicap from Howey's Leyland at 97.46 m.p.h.

After its excellent showing at Easter the Isotta was expected to be at Brooklands for Whitsun and it duly turned up, its black radiator and wheels contrasting with the bright red body. After taking third place behind Thomas and Cook in the Gold Vase Handicap, which necessitated lapping at 88.44 and 110.19 m.p.h., Le Champion proceeded to win the 'Lightning Long' from Cook's T.T. Vauxhall, at 102.5 m.p.h., the car going round really quickly, at lap-speeds of 90.99, 111.42 and 107.57 m.p.h. In June 1923 Le Champion drove all the way up to Saltburn for the speed trials. He was matched against Cook's Vauxhall, the Isotta's old rival. Cook made three runs, his best at 107.5 m.p.h. Le Champion did only two, clocking 100.76 and then 110.74 m.p.h., a very high speed for a sand course. He made f.t.d. in 20.2 sec, the Vauxhall being 0.6 sec. slower, a flying start being given, and his handling of the great chain-driven monster was described as masterly. Arriving back at the Track for the Summer meeting, the Isotta was not on form, retiring from the 'Lightning Short' after a s.s. lap at 81.68 m.p.h. and being unplaced in its other two races, although lapping, respectively, at 77.26, 100.61 and 88.15 m.p.h., and at 84.46, 101.23 and 93.44 m.p.h. It was on this occa-

The Isotta-Maybach returning to the Paddock at Brooklands, with the steep Members' banking in the background. Note the white-coated B.A.R.C. official.

sion that Eldridge was trying out his new giant Fiat.

After riding with Howey in his Leyland at a small Brooklands meeting, perhaps to see what the opposition was like, Le Champion brought out the Isotta-Maybach for the August meeting. It was in fine fettle, lapping at 71.84 and 99.41 m.p.h. in the '100 Short', but to no purpose except maybe to preserve a reasonable handicap, for it came home an impressive winner of the 'Lightning Short', from the 16 sec. mark, keeping both Leylands at bay, its speed 100.5 m.p.h. and its laps being done at 92.94 and 109.7 m.p.h. This earned a five sec. re-handicap in the '100 Long', so lap speeds were kept down to 92.41, 105.97 and 103.76 m.p.h., for a fifth place. Le Champion still had the 'Lightning Long' to run. From a limit start he was put back to 15 sec. He got off more slowly than usual, at 91.35 m.p.h. for the opening lap, but thereafter 'poured on the coals', doing 106.5 and 109.94 m.p.h., by which time Howey's Leyland was coming up fast. As they turned into the straight and made for the 'Lightning' finishing-line, Howey was right on the Isotta's tail, having done his flying lap some 5 m.p.h. faster, and it looked like a dead heat. In fact, the bigger car was fractionally ahead in this exciting finish, winning at 101.5 m.p.h. That ended 1923 on a high note for Le Champion, for he had no further victories, non-starting at the postponed (due to rain) Autumn Brooklands meeting.

The old car was out again in 1924, now entered for the first time as the

Isotta-Maybach. At the Preston & Dist. M.C. and L.C.C. Brooklands meeting it was the only car taking part. It was matched against the World's fastest motorcycle, Temple's 996 cc Montgomery-Anzani. What was intended to be a 3-lap handicap fizzled out when the car, having given the motorcycle 16 sec. start, burst a tyre on the second lap. This was followed by a 2-lap scratch race in which Temple was able to hold a lower line on the banking than Le Champion, who was troubled by a cross-wind and couldn't open his huge engine right out, so that the two-wheeler won easily. Incidentally, the new silencer arrangements caused the Isotta to appear at first with a receiver described as 'looking like an old oil drum keeping its tail up' but later it acquired proper Brooklands 'cans', as can be seen in an accompanying photograph. The Motometer on the radiator cap had been replaced by a long steam vent-pipe. It was, however, successful at the Brooklands Easter meeting, winning the Lightning Short Handicap for the Founders' Gold Cup in convincing fashion. Indeed, Le Champion made the fastest race lap of the car's career, at 114.75 m.p.h., to keep ahead of his friend Eldridge in the giant Fiat, which had given the Isotta a start of 20 sec. He averaged 104.25 m.p.h., and the 27-litre Higham Special was third—a veritable race of giants! Not surprisingly, the Isotta was put back to scratch for the '100 Long', in which it went round at 106.42 and 110.68 m.p.h. without being placed. But its 114.75 m.p.h. lap was 1.3 m.p.h. faster than the best race-lap of 'Chitty I'.

Kaye Don drove the Isotta at the 1924 B.A.R.C. Whitsun meeting but

The Isotta-Maybach in short-tailed guise, with the 'isolated' radiator.

got off badly in the Gold Vase race, doing his opening lap at only 69.66 m.p.h., so that a subsequent lap at 110.92 m.p.h. was to no avail. In the 'Lightning Short' Don got the hang of it, lapping at 90.06 and 104.85 m.p.h., although he was unplaced. But he continued to drive the car, because its owner was seriously ill in a Surbiton nursing home. In the Summer 'Lightning Short' Don lapped at 81.64 and 105.52 m.p.h. but the car was not going properly. Le Champion's illness precluded entry for the August racing, although D.W.R. Gedge was to have driven it at the Autumn meeting, but scratched from the 'Lightning Short' and retired after a s.s. lap at only 77.81 m.p.h. in the 'Lightning Long'.

Le Champion was back on the scene for 1925, and had painted the Isotta-Maybach yellow, but it was in trouble long before the Easter racing at Brooklands was abandoned because of rain. However, this driver entered his recently-acquired G.N. and the Isotta-Maybach for a Surbiton M.C. day at the Track, and in the latter, after finishing in third place in a 'Lightning Short' race, won the Surrey Lightning Long Handicap at 99.41 m.p.h. from Thomas in the Leyland-Thomas and Howey in the Lanchester. At Whitsun Le Champion entered for the Private Competitors' Handicap at Brooklands, to show he had no trade connections. The previous year he had written an article showing that it was possible to have a season's racing at Brooklands for an outlay of about £200, which is presumably roughly what the Isotta-Maybach had been costing him. The big yellow car gave Howey's Leyland 4 sec. start on this occasion, and when the situation was seen to be hopeless, Le Champion sportingly cut-out early in the race and waved Howey through; lap speeds of 90.39 and 102.48 m.p.h. proved inadequate. He then did 88.94 and 107.1 m.p.h. in the Gold Vase Handicap, but was not in the first three, while in the Gold Cup Handicap laps at 89.14 and 111.17 m.p.h. merely earned him a good fourth place.

That seems to have been the car's last competition appearance, for although it was entered for the West Essex M.C. meeting at which Thomas won his famous duel with Eldridge's Fiat, it either non-started or was unplaced. Le Champion had taken over the Fiat 'Mephistopheles' from Eldridge, and by the summer of 1925 the Isotta had been sold to Dudley Watt, who, wearing a fine black and yellow helmet, raced an aged SE5A aeroplane, which he afterwards converted at Brooklands into the DW1. It is understandable that Le Champion should graduate to the big Fiat, although it never matched-up to his Isotta and of which Eldridge may have considered that it had served its purpose after gaining him the title of the fastest-driver-in-the-World, and fame at the newly-opened Montlhéry Track. But it is remarkable that the Isotta never raced again, for the old cars had another six seasons in which to appear before they were banned by the safety-conscious Brooklands officials.

However, Watt apparently never raced the car, and the next I heard was that it had been acquired by two American couriers, who 'thought it might come in useful'. When they started to build a house near Farnham they wanted to get the old car out of the way and offered it to a Mr Sargent, who, as the local Council surveyor, had been of some assistance to them. This gentleman, who had been a Douglas dispatch-rider during the war, accepted this offer of the Isotta-Maybach, but its enormous bursts of acceleration were startling after the staid progression of his Triumph motorcycle and Morris-Cowley car, and he decided that this was no vehicle for a respectable civil servant! He drove it, still in dirty yellow paintwork, as far as a yard at Wrecclesham where his Morris was serviced. There it languished for some years before the war again engulfed the motoring scene, and was, it seems, gradually picked apart by vandals.

8. The Fiat 'Mephistopheles'

Fiat had a particularly successful road racing season in 1907, including victory in the French Grand Prix, which Felice Nazzaro had won at 70.5 m.p.h. with a four-cylinder 180×160 mm. car. It was, no doubt, these successes that caused Sir George Abercromby, Bt., a Guards officer, who lived in Aberdeenshire, to place an order for a special Fiat that would be capable of 116 m.p.h. and be able to win the 1908 Montagu Cup Race at Brooklands. The agreed price was £2,500.

This car was duly put in hand. Not a great deal of technical information is available about this famous Fiat which arrived in England during the summer of 1908 and, driven by the great Nazzaro, won the Match Race for 250 sovs. a side against the challenger, Newton driving S.F. Edge's 90 hp Napier 'Samson'. Nazzaro averaged 94.75 m.p.h. for the 27.25 miles and the Fiat's fastest lap was at 107.98 m.p.h. by hand-timing, 121.64 m.p.h. by the B.A.R.C.'s electrical timing apparatus—to this day controversy rages as to which was correct, but Nazzaro had accomplished what was expected of him and returned to Italy richer by £700 (his fee was £500 and the English Fiat Company gave him a present of £200), an enormous sum by today's values and in terms of current starting money.

The Fiat, commonly referred to as 'Mephistopheles', had a four-cylinder engine rated at 89.5 hp. The actual dimensions are obscure. The cylinder bore was 190 mm., but *The Motor* quotes the stroke as 'probably' 160 mm. Laurence Pomeroy has given it as 140 mm. and in 1913, at Brooklands, it was officially published as 185 mm., but, in 1922, after it had been idle for many years Duff presumably stripped the engine down before racing it and he certainly had the 'pots' off when installing lightweight pistons, so he had an opportunity of measuring it. Thus I accept these dimensions, which give a capacity of 18,155 cc.

The cylinders were cast in pairs, with, according to *The Motor*, two exhaust valves per cylinder, actuated by a common push-rod on the nearside, and one inlet valve per cylinder on the offside, the rockers across the heads clearing one another by a very narrow margin. The camshaft was in the base chamber on the offside, driven by exposed timing pinions, from which another spur gear drove the magneto of the low-tension ignition system. The valves were in cages some 4 in. in diameter. Cooling was by pump, but there was no fan, and the honeycomb radiator was of typical Fiat shape. The lubrication system incorporated a dashboard oil-box and belt-driven dredger pump.

The chassis followed Fiat GP design, the drive going via a plate clutch, and final drive being by side chains. The side-members were undrilled, suspension was by half-elliptic springs, and there was no body apart

The great Italian Grand Prix 'ace', Felice Nazzaro, on the Fiat 'Mephistopheles', with which he won the much-publicized Match Race against S.F. Edge's Napier 'Samson' driven by Frank Newton at Brooklands in 1908.

from two bucket seats, the long steering column being unsupported. The external gear and brake levers and the driving sprockets were drilled for lightness (or smart appearance) and the r.h. pedal applied a metal-to-metal shoe brake on the front of the gearbox and another brake on the off side of the differential shaft. Artillery 12-spoke wooden wheels were used, those at the back having heavy ornamental spokes. A big exhaust pipe fell away to a huge tail pipe on the nearside of the engine. The only attempt at streamlining was a leather fairing between the brief scuttle and the seats, and an undershield beneath the engine. *The Autocar* mentions provision for water-cooling of the back tyres from a tank on the car. The Fiat was said to develop 175 hp, against 130 hp of the 16.2-litre GP Fiats, and weighed 1 ton 9 cwt. 66 lb. Nazzaro, unlike Newton, carried a riding mechanic.

It might have been thought that Sir George Abercromby, who drove a smaller 58 hp Fiat at the Track (his colours were navy-blue coat and sleeves, red cap) would have been delighted with 'Mephistopheles', which beat Edge's six-cylinder shaft-drive Napier, and the 120 m.p.h. lap of which was loudly acclaimed. Not a bit of it! He made it known soon after the Match Race that he did not intend to take over the car, on which he seems to have paid a deposit of over £800, and he did not enter it for the Montagu Cup Race, for which he had specifically ordered it. This put Mr D'Arcy Baker, Manager of Fiat in this country, in something of a predicament. He had placed the order for 'Mephistopheles' and stood by it when Turin insisted that they must make two identical cars. There was less demand for racing cars in 1908 than when Brooklands had opened, he argued, and the Fiat Company would be lucky to sell the car for £1,000. There had been the expense of bringing Nazzaro and his mechanics over to Long Acre to tune up the car, the engine of which had

had to be completely stripped on the eve of the Match Race. D'Arcy Baker felt he had no option but to sue.

The case was heard before Mr Justice Darling at the end of the year, and the big Fiat meanwhile languished under a dustsheet. Whether doubt over the timing influenced Sir George I don't know, but certainly he disliked the idea of Fiat having a second car identical to his own, which, if Nazzaro chose to drive it, could easily beat an amateur driver.

A settlement was reached whereby Sir George accepted the Fiat for £1,250, both parties paying their own costs and withdrawing all insinuations in view of the fact that the old car was proclaimed by Fiat to be in excellent condition after Nazzaro's race. During 1909 this Guards officer had raced 8.9 hp Sizaire-Naudin, 36 hp Crossley and 40 hp Napier cars at the track and in 1910 he appeared at Brooklands with 'Mephistopheles', the car looking just as it did when Nazzaro drove it, even to the leather side valances. In the 1910 May Handicap he was on the scratch mark in 'Mephistopheles' and did his first flying lap at 106.38 m.p.h. and now, on its reappearance, it lapped very close to its best hand-timed speed of 1908. It ran in a sprint race and was out again, still on scratch, at the Midsummer meeting, when interest centred round the Fiat, in which it allowed any other entrant 62 seconds start. Its lap speed was beaten by Stirling's 60 hp Brasier (102.22 m.p.h.) and by Wildegose in the 60 hp Itala—later owned by Cecil Clutton—(101.80 m.p.h.), so Sir George was probably displeased. In subsequent races he ran his Napier, before returning to Scotland for the shooting. It seems likely that Sir George had initiated this last race in the hope of ridding himself of a liability.

In 1911 Mrs Macklin was billed to attack the World's half-mile record at Brooklands with the old Fiat but it only managed 90.94 m.p.h. Otherwise, for two years 'Mephistopheles', and indeed Fiats of any sort, were absent from Brooklands, but at the 1913 Whitsun meeting C.R. Engley appeared with the car which had achieved so much acclaim in Nazzaro's hands half a dozen years earlier. At least, everything points to this car being 'Mephistopheles', although, as stated earlier, the engine dimensions declared come out to 20.9-litres. It appears that Sir George had been posted abroad and returned his Napier to S.F. Edge, selling the Fiat to Noel Macklin, later of Invicta and Railton fame, who presumably sold it to Engley. Early in trouble in the 10th 100 m.p.h. Long Handicap, from the scratch mark, after the car had been pushed ignominiously to get it going, Engley non-started in both his races at the Midsummer meeting, but managed a lap at 93.09 m.p.h. on August Bank Holiday and laps of 98.43 and 98.62 m.p.h. at the Autumn meeting. During the abbreviated 1914 season there was no further sign of the big Fiat.

Incidentally, attempting to identify Engley's Fiat brought up a typical instance of the pitfalls a motoring historian has to face. *The Autocar* at the time remarked first of all that they believed it had 'the same dimensions as the Fiat driven by Nazzaro himself'—they were making the point that it was the largest engine ever seen at Brooklands, with the exception of the K. Lee Guinness Darracq. Even at the 'inflated' size of 20,981 cc this was incorrect; Hemery's 1911 Benz was of 21,504 cc. In a later issue the car is quite definitely described as the 1907 Nazzaro car, but earlier it was described as the Fiat 'whose previous owners were Sir G. Abercromby, Baker White, and Noel Macklin'. Baker White seems to have had the second of these Fiats, while if Macklin also owned 'Mephistopheles', the Fiat he actually raced at Brooklands for Paul Mayer was a 1906 or 1907 GP car. Just to make matters more difficult, Sir G. Abercromby, Bt., told me he thought Macklin put the engine into a boat, but perhaps it was Macklin's GP Fiat that suffered that fate.

After the interlude for war, John Duff, later to make his name with a 3-litre Bentley, discovered 'Mephistopheles' in a Fulham mews garage. At the time he was racing an S61 Fiat at Brooklands but it was getting rather long in the tooth so, encouraged by Major Cooper, who was a friend of Count Zborowski, he bought 'Mephistopheles', it is said for £100, and proceeded to get it into trim. It still retained its wooden wheels but Duff gave it a radiator cowl and streamlined tail, and a body with a proper scuttle, while I think a different, rounded-top radiator was fitted, and a revised exhaust system with tail-pipe was made up. Unfortunately, when I interviewed Duff at his race-horse training establishment near Newmarket some years before his fatal riding accident he did not remember details. The old car was repainted black, with a white body.

Duff and Cooper shared the entry of the Fiat at the 1921 Whitsun meeting, where it was handicapped to lap at 102.69 m.p.h. but failed to start. Duff also non-started in the big car's first race at the next meeting, and in the '100 Long' did a standing lap at 89.78 m.p.h., and flying laps at 106.88 and 100.82 m.p.h., but was unplaced from scratch. At the August meeting the Fiat finished second to the 1912 Lorraine-Dietrich in the 'Lightning Short', with a s.s. lap at 91.52 m.p.h. and a flying lap at 108.27 m.p.h., but after a s.s. lap at 91.17 m.p.h. retired from the 'Long' when

John Duff, better known for his Bentley records, with the rebuilt Fiat 'Mephistopheles' at Brooklands in 1922, after he had found the old warrior derelict in a London garage.

the pistons cracked and damaged the engine.

Duff did not run the big Fiat again that year but, having sold his smaller Fiat, concentrated on 'Mephistopheles' during 1922, Harry Ricardo installing light-alloy pistons in an endeavour to increase engine speed, while wire wheels were now fitted. The car was ready by the Whitsun meeting of 1922 and lapped at 89.78 and 107.10 m.p.h. in the 'Lightning Short', being caught by Zborowski's 'Chitty-Bang-Bang'. In the '100 Short' the Fiat finished third (lap speeds: 90.63 and 109.46 m.p.h.) only the Viper lapping faster, and that car was slower away. However, Duff retired sensationally in the 'Lightning Long', before completing a lap, when the new pistons proved too much for the old cylinder blocks and aged crankcase (about which Ricardo had warned Duff) and the rear pair came out through the bonnet, hung over the heads of Duff and L.G. Callingham of Shell, his passenger, then crashed back into place! Duff managed to pull up safely but it seemed as if 'Mephistopheles' would henceforth be merely a legend.

On the contrary, the car went on to achieve even greater fame. E.A.D. Eldridge, who had been racing the old Isotta-Fraschini at Brooklands with a 20.4-litre side-valve Maybach aero-engine installed, sought something faster. He had found a 300 hp six-cylinder, 24-valve, o.h.-camshaft Fiat A-12 BIS war-time engine in the Regent's Park area, after a fruitless search at the war surplus park at Slough—a new engine still in its packing case. Eldridge decided that Duff's stricken 'Mephistopheles' was just the chassis for this engine and purchased it for £25.

Going into the premises of T.H. Mallet in Hampstead to get some work

A 300 hp overhead camshaft Fiat A-12 BIS aviation engine of the type which Eldridge installed in the lengthened chassis of the aged Fiat.

done Eldridge was told by Mr Brownridge, who worked there, that he was willing to leave and help with the Fiat project. Moreover, this small works was available, so Eldridge took over the premises as well as the mechanic!

Although Jim Ames had been badly hurt when he was a passenger in a BE2c aeroplane that Eldridge had crashed at Brooklands in September 1922, he also worked on the Fiat.

A new clutch was designed, with automatic lubrication of plates and thrust-race every time the pedal was depressed, Ames making the plates (some 65 of them) at the premises of Ewart's, the geyser* people. I believe the body, too, was made there. At a Hammersmith breakers a suitable flywheel had been found in the form of a solid forging that had to be machined down. New induction pipes and a special radiator, not unlike an enlarged Fiat 501 radiator, were also specially made for the aero-engined Fiat. Originally the engine had the usual two sparking plugs per cylinder. Two more were added, making 24 in all, coil fed. An oil tank was fitted at the rear of the chassis, two scavenge pumps clearing the dry sump. The 1907 gearbox and side chains were retained, but certain parts, such as a new mainshaft, were made for it. Two Fiat double-choke, single-float-chamber carburettors were used, on the off side.

To accommodate this much longer 160×180 mm. (21,714 cc) engine the chassis was lengthened by approximately 17 in. and an elaborate sub-frame was made by Brownridge from steel plate on which to mount it. There is, I am told, no truth in the rumour that bits of London General Omnibus Company bus chassis were employed to extend the side members—proper rolled-channel to match the taper of the original side-members was used, although I believe this material may have been obtained from the well-known transport source.

While work was proceeding on the 300 hp Fiat, Eldridge raced the ex-Duff 10-litre Fiat. This proved so much trouble to maintain that Brownridge became depressed with never getting an evening with his wife, and left to work at Cookham on the 200 Mile Race Derbys.

The rejuvenated Fiat 'Mephistopheles', developing about 300 b.h.p. at 1,400 r.p.m. and 320 b.h.p. at 1,800 r.p.m., or possibly more with its new carburettors, was entered for the 1923 Whitsun B.A.R.C. meeting, the narrow two-seater body with short tail being declared as red, the wheels black. In fact, the resuscitated 'Mephistopheles' was not ready to appear and at the Essex M.C. Meeting in early June Eldridge had to be content with a couple of 'places' in his 10-litre Fiat.

However, by the Summer Meeting on 23 June the great car appeared in the 100 m.p.h. Short Handicap, running with two small bucket seats over the back axle, for there had been no time to fit the body. Eldridge had a hectic time before the race with the rear shock-absorbers breaking away from their brackets, steering wobbles and tyre trouble. The big car accelerated away impressively, but its standing lap, at 88.77 m.p.h., was slower than Thomas's Leyland, Cook's T.T. Vauxhall, Barclay's Ballot and Clement's Bentley, and a damaged rear tyre caused Eldridge to pull in. The crew worked hard to reattach the rear shock-absorbers, missing the 'Lightning Short' but coming out for the '100 Long', when the Fiat

*An instant gas water heater.

was beaten on lap speed only fractionally by the Bentley but comfortably by Thomas on the opening lap, which it accomplished at 91.17 m.p.h., shedding bits as it went, before retiring. They actually had it ready again for the 'Lightning Long' but after a s.s. lap at 85.22 m.p.h. Eldridge gave 'Mephistopheles' its way and came in.

The 10-litre Fiat had been sold to Philip Rampon and Eldridge did not appear at the August meeting, nor again at Brooklands during the 1923 season, except for a record attempt at the end of the year, when his great car successfully took the Class J s.s. half-mile record at 78.36 m.p.h. and, driving the reverse way of the track, the World's half-mile record at 77.68 m.p.h. Son of a south-east London financier, Eldridge has been dubbed a wild man, but apparently he prudently sought to cure the giant Fiat of its teething troubles before running it again, and this certainly paid dividends in 1924.

At the Easter Brooklands meeting 'Mephistopheles' started in the Founder's Gold Cup Race, on scratch with Zborowski's even-bigger Higham Special, was 6.4 seconds quicker to the Fork than that car, completed the next lap at 122.37 m.p.h., fastest of them all, to finish second behind Le Champion's Isotta-Maybach which had received 20 seconds start—a race of the giants indeed! Put back to 'owes nine seconds' in the Lightning Long Handicap, the black Fiat knocked $\frac{1}{6}$ of a second off its standing lap time, did its next circuit at 122.07 m.p.h., then slowed to 94.33 m.p.h., the task of catching Malcolm Campbell's 4.7-litre Sunbeam that had a 37-second start being hopeless. Moreover, the Fiat skidded fearsomely when leaving the banking and finished with one of its tyres actually burning. It had left Howey's Leyland as if it were stationary.

One of Ernest Eldridge's greatest exploits. With the Fiat 'Mephistopheles', after he had repowered it with a 21.7-litre Fiat aviation engine, he took the Land Speed Record to 146.01 m.p.h. at Arpajon, outside Paris, in 1924, the last time this absolute record was broken on an ordinary road.

Encouraged, Eldridge entered his black giant for the Whitsun races. In the '100 Short' he lapped at 105.29 and 118.02 m.p.h. from his customary scratch start, but was unplaced. He then drove superbly in the 'Lightning Short', lapping at 107.10 and 123.89 m.p.h. to finish second to Brocklebank's Peugeot that had the advantage of 32 seconds start. In a later race Toop was killed in this Peugeot and racing was abandoned, so the Fiat didn't run in the Lightning Long Handicap.

Eldridge had a strong liking for Paris, and consequently the Arpajon Speed Trials and the opening of Montlhéry track lured him away from Brooklands and, although entered for the Summer and August Meetings, the Fiat, now known universally as the 300 hp Fiat, did not appear. Instead it was taken to Arpajon, where it achieved the most outstanding success of its career.

Along the narrow, tree-flanked road 18 miles from Paris (today obliterated by the new autoroute) Eldridge, carrying the intrepid Ames as passenger, clocked 146.8 m.p.h. over the flying kilometre during the A.C. de France Speed Trials of Sunday 6 July. However, René Thomas, who had recorded 143.26 m.p.h. for the kilo, 143.312 m.p.h. for the mile in the 10.5-litre V12 Delage, protested that the Fiat had no reverse gear, which was called for in International regulations. So Eldridge was disqualified.

He decided that as the Land Speed Record was so nearly in his grasp he would fit a reverse gear and try again. Legend has it that he contrived this by crossing the driving chains of the Fiat but Jim Ames told me that, in fact, a reverse gear was made up in one of the many obliging little workshops that abound on the outskirts of Paris. This work appears to have occupied two days and nights, after which Eldridge returned to

The nearside of the engine in the rebuilt Fiat 'Mephistopheles', showing the six separate cylinders, exhaust pipe, and overhead valves.

Arpajon on the following Saturday, 12 July, at daybreak, and, over a road not closed officially, after slow runs when carburation was poor in the cold air, finally set the World two-way mile record to 145.89 m.p.h. and the World two-way kilometre record to 146.01 m.p.h., the last time the fastest-of-all-records was established over an ordinary road.

On this occasion Gedge rode as the intrepid passenger and must have had his money's worth, for on one run the offside rear tyre tread flew off, causing the giant Fiat to snake a little. Gedge knew nothing of this at that time, but one wonders what he felt like when Eldridge insisted on making his return run on the same tyre? An attack on the standing-mile record was abandoned on account of trouble with the electrical timing gear, but the s.s. kilo record was taken at 85.48 m.p.h. The Fiat ran on Rapson tyres and was in Brooklands trim with uncowled radiator.

There is a splendid story that the Delage, René Thomas's protest upheld, was on display in the Delage showrooms in Paris with a notice proclaiming it to be the fastest car in the world. When the Fiat broke this coveted record it is said to have been taken to Paris, and labelled in like manner, the two cars facing one another across the Champs Elysees.

Incidentally, it seems that at some stage at Arpajon, Eldridge had an oxygen cylinder on board, from which to invigorate the old airship engine. Ames tells me he found it a frightening task to hang on, pump up pressure in the fuel tank and turn on the oxygen, all at the same time but, encouraged by yells from Eldridge, managed to do so. No perceptible increase in speed was noticed! Eldridge appears to have had a love of such experiments, for I believe he had earlier bought a Warwick 3-wheeler with the intention of running it on acetylene.

It is particularly interesting, in view of the 'crossed chains' legend, that today 'Mephistopheles' has no reverse gear. Peter Gresham, who has

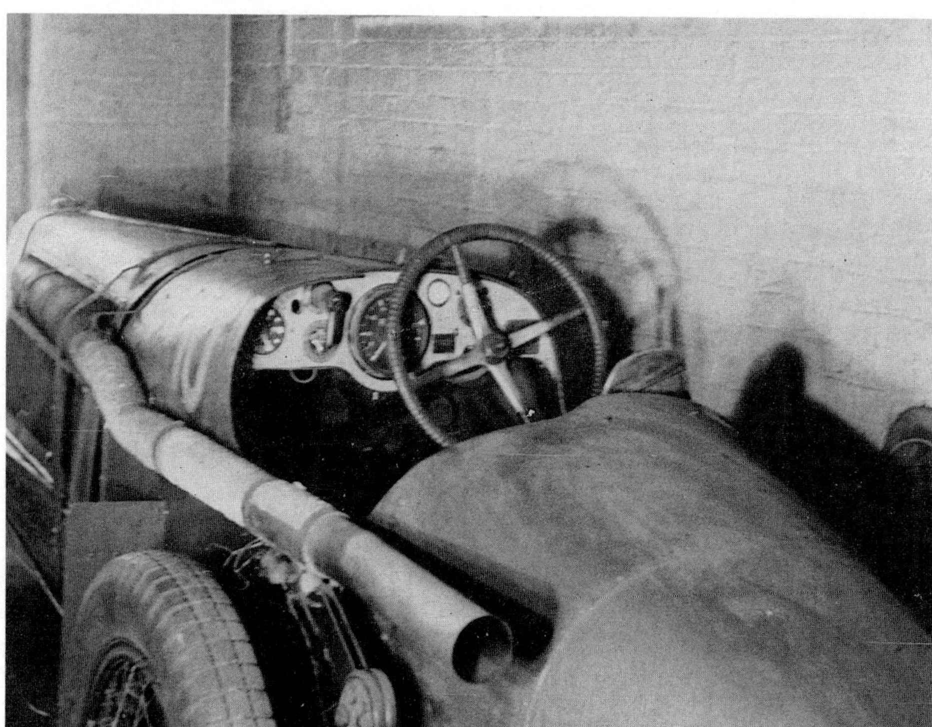

The cockpit of the immense Fiat, photographed when Peter Wike had it—and drove it on the road.

driven the old car so ably in recent Historic Racing Car events, tells me that there is no indication that one was ever fitted. The gearbox is separate from the engine but has an integral differential, and is a two-shaft box, the shafts being side-by-side, the output shaft being off-set and terminating in the differential. In Mr Gresham's opinion there were three methods open to Eldridge. He could change the crown wheel to the other side of the pinion. This would take a long time to change back even if it satisfied the officials. He could fit a third shaft above the gearbox, its gears mating with the input and output shaft gears. This set-up could have been fitted to the bolt holes of the gearbox cover, afterwards being removed without trace, but to manufacture two gears would take a long time, and what of keeping oil in the box? He could cross the driving chains, arguing that he had a reverse, but not on the timed runs. I leave you to decide what, in fact, was done.

It had long been an ambition of Brookland's enthusiasts that Ernest Eldridge should meet J.G. Parry Thomas in a Fiat/Leyland-Thomas duel, and a race of sorts, between Thomas, Duller (Bugatti) and the Fiat had been run one moonlit Thursday evening in 1923, Thomas starting 30 sec. after the Bugatti, 5 sec. after Eldridge, to win by about 30 yards from Duller, the Fiat close up and only the exhaust flames of the cars visible to those looking on. The real race nearly came about at the 1924 Autumn meeting, but Eldridge drove the Fiat over an obstruction, dragging off its oil-pump, and was again a non-starter. A lap speed of 124.33 m.p.h. had been achieved at Brooklands during the summer, breaking the lap record, although this speed was never officially recognized, whereas Thomas's later 128.36 m.p.h. was, making a match race all the more desirable.

In October, Montlhéry track, near Paris opened (its full story is told in

The thrilling Match Race between Eldridge in the Fiat 'Mephistopheles' and J.G. Parry Thomas in his Leyland Thomas, at Brooklands in 1925. After many excitements, Thomas won by a narrow margin.

my book in the Montagu Motor Book series, Cassell, 1961) and the Fiat took part in a six-lap match race against Thomas's Leyland-Thomas and a big D'Aoust. Eldridge started last, 15 seconds after Thomas. Both drivers lost offside rear tyre treads and Thomas the whole tyre, so the Fiat, which had a new cowl enclosing radiator and dumbirons, won by some 200 yards, at 121.04 m.p.h. At a later meeting Eldridge tried to improve on his own 146 m.p.h. kilometre record but was unsuccessful; in November, however, he broke Thomas's f.s. 10-mile record at 121.38 m.p.h., although unable to beat the 5 kilo, 5 mile or 10 kilo records. An offside rear tread again flew off and three laps later the tyre burst, the 300 hp Fiat really being too fast for the tyres of those days.

In December, 1924, however, the 5 kilo record was taken at 128.53 m.p.h., and early in 1925 at Montlhéry, Eldridge set this to 129.23 m.p.h., the 5 miles to 128.20 m.p.h., and the 10 kilo record to 128.34 m.p.h. At the 1925 opening meeting at Montlhéry the Fiat engaged in another match race against the Leyland-Thomas and a Borgenschutz but after four laps its 'Achilles' heel', the offside rear tyre, burst, allowing Thomas to win. In June, at Montlhéry, Eldridge indulged in yet another match race, this time against Divo in the big V12 Delage and the Leyland-Thomas; the Fiat finished last.

The car now returned to England, but I think probably this was because had it remained in Paris, customs regulations would have been broken and much money lost. At all events, Mr Ames tells me how he suddenly remembered that the car was due back, the batteries were hastily charged, and they set off for Dieppe in pouring rain, the undershield dragging on the road, Ames burning his overcoat on the exhaust pipe but the boat being caught in the nick of time.

Having got it back to Brooklands, Eldridge decided that the long-awaited Match Race with Thomas might as well take place. It was duly contested, for stakes of £500 a side. The race proved one of the most exciting ever seen at the Track, and devastatingly dangerous. Eldridge, carrying a passenger and bare-headed, scorning goggles over his glasses, led by some 200 yards as they went on to the Byfleet banking, and the Fiat was still in front as it slewed over the Fork in a series of wild swerves. The race was over three laps and Eldridge still led at the end of the second round, taking the Byfleet banking in a cloud of dust, the start having been from the half-mile box on the Railway straight. However, Thomas's superior suspension and his track-craft enabled the Leyland-Thomas to close up and, Eldridge sportingly keeping low on the bankings, the Welshman got past at the Fork as Eldridge had to lift off momentarily. Once again the Fiat proved too heavy and fast for its tyres, the offside rear tread flying to the tree tops as it ran onto the Members' banking. Thomas lost a front tread but both men—they were tough in the 'twenties'—kept going, Thomas winning at 123.23 m.p.h. to the Fiat's 121.19 m.p.h. The Leyland-Thomas lapped at 129.70 m.p.h., the Fiat at 125.45 m.p.h., but neither was recognized officially. A contemporary account of the race brings to light vividly the excitement an aero-engined car brought to the onlookers when it was driven really fast:

> The cars went off together from the mile box on the Railway straight, Thomas, to the astonishment of the crowd, actually taking the lead. Before the end of half-a-mile Eldridge's huge machine fairly hurled

itself past the Leyland, and took the banking in one long skid. Still accelerating, Eldridge got well away, then gave the more experienced people the shock of their lives, and the inexperienced sheer delight, by coming wildly off the banking in one series of awful swerves, the Fiat sliding about until it seemed impossible that the driver could hold it.

Hold it he did, however, and still swinging about, flew out of sight behind the hill. The knot of spectators at the Fork hurriedly took cover, several people refused to look, and then the car roared past again, this time more steadily. In the meantime Thomas was really travelling, and slowly the Leyland gained. Once Thomas tried to pass, only to slide down the Banking. Then both cars entered the Railway straight on the last lap. A huge cloud of dust flew up as the Fiat ran sideways onto the Byfleet banking and skidded again coming off the Byfleet banking to the Fork. Then Eldridge deliberately cut out, to let Thomas pass if he could. Like a streak the Leyland passed and held its lead.

Eldridge had set up residence in Paris and was deeply immersed in racing his 1.5-litre Anzani-engined Eldridge Specials, so in July 1925 he sold 'Mephistopheles' to his friend L.C.G.M. Le Champion, who had been so successfully racing the Isotta-Maybach, which he used to skid round in circles on the aerodrome until the tyres burst, when that car was due to be reshod. This wild-man flew wartime aeroplanes and wrote a book of psychic short stories, entitled *Foretold,* under the nom-de-plume of 'Streamline' in 1934.

The Fiat ran in the Match Race sans radiator cowl and on Dunlop tyres, and in this form Le Champion took it over, repainting it green. He kept it for a time at Temple Street Garage, in Rugby (which still exists), and in that area they still talk of his giant racing cars storming the roads.

At this time Betty Haig remembered having several runs, at the age of 18 or 19, in the big Fiat, when it was being tested on a straight, none too wide, road near Daventry. It was inexpedient to get out of third gear and at full speed it needed most of the width of the road. She recalled that a new, shining Delage saloon on French number-plates that had just come up from Monte Carlo was used on one occasion to tow-start the Fiat's engine, which when it fired had a bad effect on the Delage which Betty was driving. She found the heat from the aero-engine exhaust agony even with her leather coat stuffed down beside her. There was also a girl who got herself into the papers after being flung from the car in October 1925, as it went over a bump on the sands at Littlestone-on-Sea, escaping with a grazed elbow and injured hand after turning several somersaults.

There was an interlude during Le Champion's ownership of the car when he took it out to Australia, possibly to race it at the Maroubra track. He kept the car in the showrooms of Garratts Ltd, at 173 Elizabeth Street, Sydney, from where he could, and probably did, drive it to the new track, a matter of about two hours in those days. Perhaps Le Champion found the big car unsuitable for the course but it could have been displayed there under the floodlights used to illuminate the racing. He had possibly hoped to sell the Fiat but, not succeeding, brought it back to England.

He brought it out at Brooklands on August Bank Holiday, 1925, lapping at 101.23 and 111.42 m.p.h. in the Lightning Short Handicap, the acceleration terrific but the new owner cautious on his first outing, giving

The Fiat, with cowled radiator as used by Eldridge at Montlhéry, in a very public setting in Sydney, Australia, where Le Champion had taken it with the unfulfilled hope of racing it and perhaps of selling it.

way to Thomas, and rather slower, at 94.15, 110.43 and 110.92 m.p.h. in the equivalent long (8.5 miles) race. He was unplaced and failed to start at the Autumn meeting because the Stewards objected to the condition of the tyres. The old Fiat, now eighteen years old, did not run again at a B.A.R.C. meeting, but Le Champion continued to use it in sand races and speed trials, notably at Blackpool in 1926, when it provided much flame and smoke but nothing much else and was on very sad tyres.

W.G.S. Wike acquired the car some years before the war and I cannot do better than print the letter he wrote me in 1988 relating to that period of the Fiat's career:

> I have beside me as I write a tattered receipt, which reads as follows:
>
> East Cliff, Preston.
> Received from W.G.S. Wike, Esq., fifty pounds (£50.00) for 300 h.p. F.I.A.T. Racing Car complete with battery and all loose parts.
> 9.x.31 Signed (over two George-V red penny stamps)
> N. W. Sharratt
>
> Before the 1914 War the Fiat had four cylinders, and was raced by John Duff. After the war it was bought by Ernest Eldridge, who scrapped the engine, lengthened the chassis, and fitted a Fiat airship engine, which could be bought from Coley's at Kingston for £25 each. These engines had two plugs per cylinder and two priming or compression taps; he removed the latter and fitted two more plugs, removed the two magnetos, and fitted four coil ignition distributors in their place. Then he and the Fiat went to Arpajon in France (near Paris) and on a

narrow country road, took the Land Speed Record. I have been to Arpajon, but nobody seemed to know just where this happened.

He then sold the car to another Brooklands character called Le Champion. He had a friend called Niel [sic] Sharratt, who was also a friend of mine. They were driving up from Le Champion's home at Leamington Spa to Preston, but 4 sets of coil ignition and no dynamo got them as far as Knutsford, where the car was put in Elstone & Knight's garage to have its battery charged, and there it remained until October 1931, when I bought it, and a friend from Preston called George Gregson, who had an engineering works, paid me £25 for a half share. We overhauled the engine (probably unnecessarily) and fitted new Marelli aero-plugs (also from Coley, 2s.6d. per dozen!) and entered it for a race meeting on Southport sands. It was quite unsuitable for sand-racing and after being soundly beaten by Percy Stephenson on a side-valve 747 cc Austin, we decided that it was possibly more use as a road car (after all, it had done its stuff at Arpajon on the road, hadn't it?). George and I had a friend called Charles Naylor, and Charles and I used the car for inspecting certain public buildings; there was of course no starter, but we could always rely on a push start from the other customers. I don't remember being bothered by the Southport Police, and we took the precaution of borrowing a pair of £5 trade-

In 1961 the famous and historic Fiat 'Mephistopheles' was restored and appeared at a Vintage Sports Car Club meeting at Oulton Park, where W.A. Briggs drove it in the Edwardian race.

The aero-engined Fiat on the line at Oulton Park with Cecil Clutton's 1908 GP Itala beside it.

plates from a friend in the trade, John Bradshaw, so as to be Legal (the annual tax was something we dared not even calculate).

George took the car back to his works at Preston, and I sold my share in it to him. There it lay until 1945. George was killed at Dunkirk but we did not like to approach his mother about the car, as she hoped against hope that he might have somehow survived. After the war George's cousin Jim took over the works, and one day Charles Naylor offered to buy the car from the estate, but Jim said, take the b****y thing away, I don't want it; next day Charles turned up with a low-loader, and away it went to a garage near his home at Knutsford owned by a friend called Peter Gresham. I was called in to explain why the mags would not give a spark, and took quite some time to persuade him that they weren't mags, but only distributors! The car ran at Oulton, Gresham at the wheel, and was then rumoured to have been bought by Fiat for their museum at Turin, for £10,000 and a Fiat 128 . . . .

In 1961 came the exceedingly pleasant surprise of finding that Mr C.E. Naylor of Manchester, who owns the Rozalex firm, had prized the famous car out of its retirement. It was beautifully rebuilt by his son John Naylor, assisted by W.A. Briggs, how drove it in the rain at the 1961 V.S.C.C. Silverstone Races, and made extremely welcome appearances at Silverstone, Prescott, Oulton Park, Shelsley-Walsh and Brighton, winning a race when driven by Gresham at the 1961 V.S.C.C. Oulton Park Meeting, at 58.25 m.p.h. It ran as well as ever on Shell petrol, Castrol oil, an Exide battery for ignition, new Reynold chains, 24 Champion plugs and Dunlop wheels and tyres.

As recounted by Peter Wike, the Fiat is now on display in the Fiat Museum at Turin, in almost exactly the form in which Eldridge and Le

Champion ran it, although the driver's seat protrudes rather more from the cockpit, possibly because the tail is lower. Another of the rare aero-powered racing cars saved! Chain guards were fitted before its appearance in V.S.C.C. races.

The 54-year-old Fiat going well in a V.S.C.C. race at Oulton Park.

10. The White Mercedes

Count Zborowski had imported from Germany a 28/29 Mercedes chassis equipped with a sketchy test body, a car with a six-cylinder overhead camshaft engine, a direct third gear in its four-speed gearbox, and somewhat rudimentary four-wheel-brakes. Besides this fast Mercedes the Count had a 1921 37.2 hp six-cylinder overhead camshaft Hispano-Suiza touring car, which he preferred to the Mercedes on account of its smoother engine and better, servo-operated four-wheel brakes. The Mercedes might be able to overtake the Hispano-Suiza up Dover Hill in its direct third speed, but the Count's affections were with the French car. It was decided, therefore, to turn the Mercedes into another aero-engined car for Continental touring. Thus 'Chitty III', sometimes called 'The White Car', was born.

The original 7.25-litre engine was taken out and found its way into a boat. The chassis was lengthened and tie-rod braced, and given the highest final-drive ratio the axle casing would accommodate. A six-cylinder Mercedes aeroplane engine of 140×160 mm. (14,778 cc), or 160 nominal horse-power—as used in Albatros D1, Aviatik CII, Fokker DVII, Pfalz DIII and other German fighters—was installed, these war-reprisals engines being obtainable at the time for about £60 unused or for £30 in reconditioned form. To improve the brakes, the Westinghouse people were persuaded to make up the parts necessary to operate them under pressure, controlled by a genuine locomotive driver's valve. At first, pressure was obtained from the rearmost cylinder of the engine, via a reducing valve, but sooting up was a major problem, so a small air-compressor was installed, after which the system worked very well,

Count Zborowski's 14.7-litre aero-engined Mercedes, the 'White Mercedes' at Brooklands in 1924 when it lapped at 112.68 m.p.h.

The 'White Mercedes' was used for Continental touring and even for venturing further afield, such as penetrating into the Sahara Desert, sometimes in the company of 'Chitty II' as a sort of baggage wagon. It is seen here on one of these expeditions.

except for delayed action. A racing body was put on and that was 'Chitty III', a comparatively modern shaft-drive car, but still a monster by the average engine-sizes of 1923.

Being deprived of 'Chitty I' for the last Brooklands meeting of 1922, Zborowski entered 'Chitty III', but a broken clutch ball-race in practice caused chronic clutch slip and the car was withdrawn.

By 1924 Zborowski was involved as a Mercedes driver in races of Grand Prix calibre, but his visits to Unterturkheim to discuss these and kindred matters were made in 'Chitty III'. This car was entered, as a Mercedes, for the Brooklands Whitsun meeting of 1924, in the Private Competitors' Handicap, indicative of the fact that Zborowski still retained his amateur status. It got round at 93.62 m.p.h. but was unplaced. However, it was merely warming up to its task, because in the subsequent 100 m.p.h. Short Handicap for the Gold Vase it lapped the Track at 104.63 m.p.h. on its fastest lap, and won by a quarter-of-a-mile, at 98.5 m.p.h., from Duller's Bugatti and the venerable 1912 Grand Prix Lorraine-Dietrich. The Count brought the car out for the Summer Brooklands Meeting, when it was limit starter in the Lightning Short Handicap, in which it finished second to Thomas's Leyland-Thomas. Zborowski drove the Mercedes again in the Lightning Long race, preferring it to his Ballot, which he entrusted to Clive Gallop. This was a very stirring race, Thomas beating the Mercedes by a mere couple of yards and breaking the absolute lap-record in doing so. Zborowski's best lap on

Another view of the Mercedes, into which Zborowski put an aero-engine of the same make, on one of his Continental forays.

Remarkably Noel and his friend Pole resurrected what was almost certainly Zborowski's 'White Mercedes' and ran it at Brooklands, including finishing in the 1929 B.R.D.C. 500 Mile race, although by then it was quoted as having a 16.2-litre aero-engine.

The ex-Noel Mercedes as it was when that great enthusiast for exciting motor cars, Clive Windsor-Richards, ran it on the road.

this occasion was at 112.68 m.p.h., and towards the end of the race he is said to have mistaken a burst silencer for bits of a rear tyre tread and eased up. That was the last occasion on which Zborowski appeared at Brooklands in one of the 'Chittys'.

The White Mercedes is thought to have passed into the hands of Noel and Pole, who raced her again at Brooklands in 1929. It seems that Howey entered her for the 1926 Easter Private Competitors' Handicap but never got to the starting line. There is no positive proof that the car Noel and Pole discovered in the British Mercedes-Benz works in London *was* 'Chitty III', and some doubt is caused because the engine was then declared as a six-cylinder of 148×170 (17,850 cc). However, it is unlikely that two such Mercedes existed and the explanation is probably that Howey had put in another, larger engine which he may have acquired with the two 'Chittys' he bought from Zborowski, or that the engine dimensions were subsequently mis-declared.

The fact remains that this ancient giant, now fitted with a fine long-tailed fabric racing body by Martin Walter of Folkestone, contrived to finish the course of the 1929 B.R.D.C. 500-mile race after being tried out in the Skegness sand races. The following year it finished second to Daybell's 30/98 Vauxhall in the Cornwall Senior Short Handicap, lapping at 110.43 m.p.h. At the end of that season such cars were banned by the Brooklands' authorities on account of possible metal fatigue.

The giant Mercedes languished for some time at Thomson & Taylor's

Brooklands premises until it took the fancy of Clive Windsor-Richards, who had raced 30/98 Vauxhalls and other fast cars at the Track. He purchased it for £50 and bravely used it on the road in the 1930s, finding it reasonably docile. He disposed of it to Lord Carlow for £75. His Lordship met Clive at his local Bentley Station in Hampshire, complete with bowler hat and rolled umbrella, and quite unconcernedly drove away in the enormous Mercedes. That was the last anyone has ever heard of it. One must presume it to have been broken up.

11. The Higham Special/Thomas Special 'Babs'

Count Zborowski had already had considerable success at Brooklands with his legendary 23-litre Maybach-engined 'Chitty-Bang-Bang I', and much fun touring English country roads and the Sahara Desert in his 19-litre Benz-engined 'Chitty-Bang-Bang II', before the 27-litre Liberty-engined Higham Special came into being. Zborowski was a wild, adventurous, romantic, fun-loving person. He may have hoped for widespread fame as a racing driver when he bought the Aston Martin firm so as to have a modern voiturette to race, or when he accepted a place in the 1924 Mercedes Grand Prix team, but at Brooklands he liked the romance and excitement of driving his giant aero-engined cars.

Consequently, when seeking something even more 'frightful' than the now-famous 'Chittys', he bade his Chief Engineer and Racing Manager, Clive Gallop, keep to what can be described as the 'Chitty Formula'. Gallop had visualized, even in 1923, when the Count discussed his next monster racing car, something a little more sophisticated than the crude if capable 'Chittys', which after all, had made-do with pre-1914 Mercedes chassis, chain-driven, back-wheel braking and other primitive mechanical features.

To such entreaties, Zborowski turned a deaf ear. More, he insisted on a simple, channel-section chassis frame, whippy as this must be, for the installation of the American Liberty aeroplane engine (No. S2) of 127×178 mm. (27,059 cc) and some 400 hp which he had acquired. If the chassis flexed, this could be met with those wire braces which were a picturesque aspect of the 'Chitty' chassis frames. Other primitive parts lying about the workshops at Higham, the Count's home near Canterbury, could be used in building up the car.

When you are employed by, and are firm friends with, a virtual millionaire, particularly such a mercurial one as Vorow Zborowski, it is not wise to argue! So Gallop went ahead, designing a crude chassis frame which Rubery Owen constructed, putting into it the bulky V12 Liberty engine, and driving this through a Mercedes scroll clutch and a gearbox from the pre-war 200 hp Benz which Zborowski had only recently given up racing on the Track, as being in too dangerous a condition, even for his adventurous tastes. The front stub-axles for the new car apparently came from a 1908 GP Mercedes, which I imagine was the car in which Zborowski's friend Hartshorne Cooper had been killed while practising at Brooklands in 1921, after he had, no doubt through the persuasion of 'Chitty's' creator, put a V8 Clerget aero-engine into it.

Gallop had put aside his visions of a chassis with deep side members, lattice boxed, and Perrot front-wheel-brakes. Final drive was by side-chains, as on the 'Chittys', with a narrow guard of $\frac{3}{8}$ in. nickel-chrome steel over the offside one; and the car, named the Higham Special, was given a Bligh two-seater racing body and a cowl over its radiator, which was somewhat wider at the top than at the base. Suspension was by half-elliptic leaf springs, and the car, which was the largest-engined Brooklands racer, ran on Rudge-Whitworth wire wheels shod with 880×120 tyres. It weighed 1 ton 13 cwt., with the back wheels carrying 8 lb. more than the front ones. Starting was accomplished either by winding on a side handle, which turned the engine through reduction gearing, or by using an electric starter energized from batteries carried on Zborowski's Mercedes service lorry.

When this great car was first tried out, the reverse gear on its short shaft broke out of the gearbox, the aluminium of which, some 17 summers ancient, showed a tensile strength of a mere 4-tons. Immediately Gallop was instructed to fly (in 1923!) to Mannheim, there to obtain, and return with, the last Blitzen Benz gearbox available.

For what purpose was the Higham Special built? Col. Gallop (who was unhappily killed many years ago while driving his Renault Dauphine near his home in Leatherhead, without any other car being involved) told me it was for the fastest Brooklands races, but also for the record attempts at Arpajon and possibly Daytona. At the time the so-called Land Speed Record stood to the credit of K. Lee Guinness's 350 hp V12 Sunbeam, which had devoured the two-way kilo at Brooklands at 133.75 m.p.h. Zborowski may well have had in mind going faster than that, at

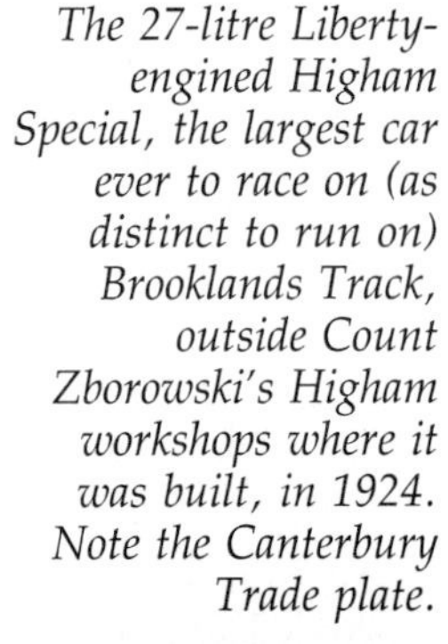

The 27-litre Liberty-engined Higham Special, the largest car ever to race on (as distinct to run on) Brooklands Track, outside Count Zborowski's Higham workshops where it was built, in 1924. Note the Canterbury Trade plate.

the 1924 Arpajon Speed Trials, where Eldridge was to set the Land Speed Record to 146.01 m.p.h. in his 21-litre Fiat, the last time this record was broken on a public road.

First, naturally, Zborowski entered the Higham Special for a Brooklands race. It failed to get away in the 100 m.p.h. Short Handicap at the 1924 Easter meeting, but the huge white car lapped at 116.91 m.p.h. in the Founders' Gold Cup race. This probably kept the Count busy but, in fact, Eldridge's big Fiat was faster by 6.4 sec. on its standing-start lap, and 6.45 m.p.h. quicker on its flying lap, and the Fiat was beaten by yet another aero-engined car, Le Champion's 20-litre Isotta-Maybach, which had given both the Fiat and the Higham Special a start of 20 sec. After this Zborowski turned to his shaft-driven Mercedes-engined 'Chitty III', and I do not think he had an opportunity to drive his biggest car again before he was killed in a 2-litre Mercedes during the 1924 Monza Grand Prix. Whether the Higham Special would have served Zborowski's purpose we shall never know; it had run quite well, being faster round Brooklands than 'Chitty I', but its frame, as Gallop had forecast, was too whippy, it tended to get too hot, but handled comfortably when doing 120 m.p.h. along the Railway Straight, at a mere 1,400 r.p.m. or so.

We now come to the second phase in the career of this car. When Zborowski's estate, which included his passenger-hauling model steam locomotives, was disposed of, Parry Thomas bought the Higham Special, it is said for £125. Thomas, after designing the luxury Leyland Eight for Leyland Motors Ltd, had gone to live inside Brooklands Track and had built up a very fine reputation as a racing motorist and record-breaker.

Higham House, as it was when all the 'Chitty-Bang-Bangs', the 'White Mercedes', and the 'Higham Special' were conceived and made in those roaring twenties.

But in spite of the money that was to be made from bonuses in the record-breakng field, I do not think Parry Thomas was a particularly wealthy man. To break the Land Speed Record would be worth a good deal financially, including Lord Wakefield's £1,000-a-year until it was rebroken, or some £100,000-a-year in terms of modern money. In any case, this was a challenge no self-respecting racing-car designer could overlook.

Thomas was a great engineer, but his Leyland Eights had been stretched about as far as they would stretch. Denis Jenkinson, in his *Racing Car Pocketbook*, quotes their top speed as 150 m.p.h. By early 1926 the ultimate in motor-car speed was 152.33 m.p.h., set up on Southport sands by Henry Segrave in the 4-litre V12 supercharged Sunbeam 'Ladybird'. But others were preparing to travel much faster—Malcolm Campbell with his Napier Lion-engined 3-ton special, and Segrave himself, with the twin-engined 1,000 hp Sunbeam, which was ultimately the first car to exceed 200 m.p.h. and which never fails to impress me with its brute-force crudeness when I see it in the National Motor Museum.

In 1926, then, Thomas had a chance. But only by finding a faster car than his famous Leyland-Thomas. He had neither time nor money to build a fabulously-expensive special car for his bid to take the Land Speed Record. I have no doubt that he could have produced such a car; but it would have needed rich backers, and I do not think Parry Thomas had the necessary money himself. It would also have disrupted his Brooklands' workshops, already busy, or about to be, with projects like the Leyland-Thomases, the Thomas Specials in 750 cc, 1,100 cc and 1,500 cc forms, the development of the 'Brooklands'-model Riley Nine, racing a Bentley, tuning an Invicta, and so on. And to build a brand new car for the Land Speed Record would have taken time, time during which

The vee-twelve-cylinder Liberty aero-engine of the Higham Special. The coil-ignition distributor can be seen at the front of the offside camshaft.

Campbell and Segrave would have had the opportunity to lift the speed even higher.

So, on several counts, the Higham Special was a sensible car to use, although privately Thomas may have wished for something less aged and more substantial. However, he had Zborowski's scarcely-used racing car brought to his Brooklands' workshops and there decided on the few modifications possible. (It is somewhere about this time that the affectionate name 'Babs' was given to the giant car, although officially it was entered as a Thomas Special. The popular story is that a mechanic, impressed by the bulk of the Liberty engine, chalked this name, or perhaps it was 'Baby', on it and that Thomas, taciturn as he was reputed to be, was amused and used the name from that day on. But more recently it has been suggested that the car was named after a relative or friend of Thomas's; perhaps a child.)

Thomas took the Higham Special in hand to the extent of fitting four Zenith carburettors to the engine, two at each end of the manifolds which were within the vee of the cylinder blocks, making his own pistons for it, and substituting a Thomas clutch for the old Mercedes scroll clutch with which Zborowski had been content, the Count having held that although it was slow to take up the drive it gripped well enough to justify its presence. Thomas also altered the front cowling to come down over the dumbirons, and generally changed the body so that it was similar to that on his Leyland-Thomas racing cars. He did not like those 1908 stub-axles and substituted a Leyland front axle. Some reports say he strengthened the chassis and changed the weight distribution, but I do not see that much could have been done in that direction.

It took him about a year to get round to these alterations, so it was 1926 before 'Babs' was used by its new owner. Thomas had taken the great car

The Higham Special as modified by Parry Thomas into the Thomas Special 'Babs', with which he was to raise the Land Speed Record to an eventual 171.02 m.p.h. at Pendine sands, only to crash fatally when trying to improve on this speed. Thomas is seen having 'Babs' refuelled in the Brooklands' Paddock.

out for a trial run round the sheds at the track in the dusk one April evening, the enormous engine starting at the second push. (I think it was always started in this fashion, Thomas having apparently dispensed with the side handle and electric starter.)

The car's first track appearance was at the 1926 Whitsun meeting. The Brooklands authorities must have welcomed it as an inveterate crowd-puller, especially as during the previous month it had broken the coveted Land Speed Record. But in its first race, the Gold Vase, Thomas did not drive it. Instead he entrusted the car to John Cobb, although as originally entered, the Thomas Special 'Babs', 127×177 mm. (26,907 cc), white and blue, as it appeared in the Race Card, was to have been driven by its owner. Cobb gave Howey's 4.8-litre straight-eight Ballot a start of 19 sec., being on scratch, and was quite out of the running, although lapping at 99.21 and 111.92 m.p.h. It seems that Thomas may have been intent on conserving his handicap, because he drove Rapson's Lanchester in place of Cobb, being likewise unplaced. Cobb, who was down to drive 'Babs' in the Lightning Short Handicap, now let the owner take the car, Cobb simply swopping cars with Thomas, who had entered his Leyland-Thomas. The big car went very well, lapping at 108.27 m.p.h., which was by far the quickest standing-start lap in the race, and then at 125.77 m.p.h., bringing Thomas into second place behind Jack Barclay's T.T. Vauxhall, which had gone off 30 sec. before him. This opening Track performance must have pleased Thomas, although in the Gold Star race 'Babs' lapped at only 99.41, 110.43, and 111.17 m.p.h., Cobb in charge, whereas Thomas, back in his Leyland, got the 7.2-litre car round at 107.24, 127.38 and 126.09 m.p.h.

Thomas brought 'Babs' out again for the August races, the nickname no longer appearing in the programme. In the '100 Short', from his customary position on scratch, he lapped at 112.68 and 123.58 m.p.h. without being placed. But in the 'Lightning Short' he came home second to Kaye Don's 5-litre Sunbeam, the big car going round at 111.92, and 120.01 m.p.h. That was 'Babs's' last appearance in a race, although Thomas won two races in his Leyland-Thomas that afternoon. Even though its horse-power was quoted as 500 or 600, depending on which journalist you believed, and it was said to get up to some 160 m.p.h. by the end of the Railway Straight, Thomas having then to apply the outside hand-brake to steady it, 'Babs' was no doubt a difficult car on Brooklands. The Leyland-Thomas was faster, giving Thomas the outright lap-record at this time, of 129.36 m.p.h., set up in 1925. But, although the smaller car was also more accelerative, 'Babs' was used that year to break World and Class A short distance records. In May she set the World's s.s. kilo and mile distances to 86.9 and 98.87 m.p.h., respectively and in mid-October she took the 5-mile and 10-mile Class A records, at 124.25 and 122.91 m.p.h., respectively from Sunbeam and Fiat, and set up a new kilo record of 123.91 m.p.h. at Brooklands, 'Babs' weighing out at 34.5 cwt.

This, however, is to anticipate, for before her first Brooklands appearance 'Babs', as I have said, had already broken the Land Speed Record, so that she was not only a great draw to the public but had accomplished what Thomas bought her to achieve. In 1925 'Babs' had been taken to Pendine, equipped with discs on the back wheels, a cover over the passenger's seat, no aero-screen, and stub exhausts. But the

October weather defeated Parry Thomas's Welsh pilgrimage.

In 1926 the car was again taken to Pendine, now with the long Brooklands' tail, and four air-intakes at the front of the bonnet which led cold air to the carburettors; to the rear pair via long metal boxes. On 27 April 'Babs' covered the two-way kilo at 169.238 m.p.h. and the two-way mile at 168.074 m.p.h., breaking Segrave's Sunbeam speeds by a handsome margin. And that was going very fast, in an ancient chain-drive car, on the tyres of those times!

But Thomas was not content, and the following day, after removing the air-intakes, streamlining taking precedence over the last ounce of power, he increased these mean speeds to 171.02 and to 170.624 m.p.h., respectively. This record bid occupied from Monday to Wednesday, the best run being at 172.331 m.p.h., on the Tuesday, when the records were first broken. Some runs were made on pump Shell, but eventually the fuel used was 60 per cent Shell aviation spirit, 40 per cent benzole. The wheels were Rudge-Whitworth, the plugs K.L.G., the driving chains were Coventry, and Laystall had made new camshafts for the engine, which had Delco coil ignition using a C.A.V. battery. Shell oil was used and the old Benz gearbox was apparently still in the chassis.

We come now to the tragic end of 'Babs' and of Parry Thomas's splendid motoring achievements, a tragedy brought into the limelight again by the 1969 Pendine controversy. By 1927 Thomas must have realized that his Land Speed Record was in jeopardy. Campbell had his new Napier-Campbell ready; Segrave was about to sail for Daytona with the twin-engined Sunbeam. Indeed, Campbell took the record to 174.883 m.p.h. for the kilo and to 174.223 m.p.h. for the mile at Pendine in February, on his third visit with this car to the Carmarthenshire beach.

Thomas, however, was ready and game. He had put a more sloping radiator on 'Babs' and almost completely cowled this in; had improved the Liberty's induction system, and had cleaned up the body contours, even to covering the driving chains with streamlined fairings. Some say he had lowered the chassis a little, but if so, the big engine still sat very tall in it. After a short run on the Track, hampered by its winter repairs, 'Babs' was again taken to Pendine sands, apparently not on a hired solid-tyred, chain-driven articulated Scammell lorry, as on previous occasions, but on a trailer, behind a lorry.

The exhaust stubs, six each side, were now flush with the bonnet-sides, discs covered both sides of the wire wheels, tower-like fairings covered the stub-axle pivots (perhaps to exclude sand?), and flat horizontal guards behind the front wheels to keep down sand and spray were used. An aero screen was crudely fitted to a wooden cross-piece in the cockpit, and the bonnet had been built up smoothly over the engine. The name 'Babs' was on the sides of the bonnet.

Thomas had intended to have a crack at Campbell's figures in February, but influenza prevented this. Still feeling ill, the 41-year-old Welsh driver was there by 1 March. The weather caused delays, but on Thursday, 3 March, the runs were on. At first the carburettors and plugs gave trouble. Then Thomas set off in earnest. He never returned. Beyond the measured mile 'Babs' slewed round, rolled over, resumed her wheels and caught fire. The fire was soon extinguished but Thomas was dead; decapitated, it has always been said, by the broken offside driving chain.

The cause of this tragic accident is still a subject for discussion. The

'Babs' just after Parry Thomas had been killed at Pendine.

generally accepted theory is that the offside back chain broke, either through encountering some object thrown up from the sand or because Thomas shut off too sharply, reversing the load on it. But there are those who consider that the offside back wheel collapsed first, one of its broken spokes then getting under the chain and severing it, a theory substantiated by Reid A. Railton, who carefully examined the wreckage.

The Dunlop Technical Department, Racing Division, has very thoughtfully provided me with data sheets relating to tyres used for Land Speed Record attempts from 1926 to 1939, from which I note that on the 1926 run 'Babs' had 33×5 Dunlop tyres of 10-ply cotton, revolving at 1,760 r.p.m. at 170 m.p.h. The centrifugal force index was 89, lowest of any of the subsequent record cars, and the weight of the Thomas Special is quoted as 35 cwt., distributed 5.5 cwt. over each front wheel, 12 cwt. over each back wheel. The power output is given as 500 hp. For some unexplained reason, no data is quoted for the unsuccessful 1927 attempt. On this occasion discs enclosed the spokes of the wheels, which were uncovered, at least on some of the runs in 1926.

Another theory, held by the late John Bolster, is that a radius-arm broke. This would certainly be likely to both throw a chain and cause a wheel to collapse. But as 'Babs' rolled over one complete turn sideways during the accident, one cannot be sure this somersault did not cause the radius-arm to break. Indeed, although Thomas was reported to have been decapitated by the broken chain, which had smashed through its guard and the body fairings, it must not be forgotten that the car slid along upside down for a short distance; this in itself could surely have caused his head injuries? On the other hand, if something went amiss at the offside rear of 'Babs', the driver would tend to look down and back, or might even have been thrown into this position, and thus he would be in the path taken by the broken chain. It has been said that, if the electrical timing apparatus had not proved temperamental, Thomas might never have started on this third run along the course, but Hugh Tours's masterly biography of Parry Thomas does not substantiate this, 'Babs' being, according to this historian, too slow rather than untimed, on her earlier runs.

The rest of this story is soon told. The wrecked car was dragged by tractor up the beach and the next day to a garage in the town. After the inquest that day, it was buried behind the sand dunes, after Thomas's staunch crew had damaged as much of it as possible. They slit the driver's leather coat and threw this into the 'grave'. They then erected a cairn of stones above the burial spot. Thomas's body was taken back to Brooklands and buried in Byfleet Church cemetery. After a time, golfers removed the cairn, and when the land where 'Babs' was buried passed out of local ownership into the hands of the Ministry of Defence it was thought that at last the car was unreachable, especially when a concrete apron was laid over most of the burial place, this being part of an incidental installation at the Proof and Experimental rocket ranges.

Then, early in 1969, Mr Owen Wyn Owen obtained permission from the military authorities to dig the car up. This gave rise to bitter controversy in Pendine, one side saying this would be sacrilege, the other that if 'Babs' could be rebuilt (which a suitable inspection committee—composed of museum experts, engineers, and racing-car historians—would have to decide should the car be dug up) she would be the finest permanent memorial the great Welsh driver and engineer could possibly have.

When I heard that 'Babs' was likely to be disinterred I went down to Pendine, where a rather droll situation had developed. At one hotel I was asked to sign a petition against this, at another hotel to sign one expressing the hope that the old car would reappear. I obtained an interview with the Commandant of the military base where 'Babs' was buried, and he told me that he was puzzled about how much controversy there was about a matter he did not fully understand, which was sometimes causing friction between soldiers and civilian workers at the site, who had previously been amiable!

I had long held the view that there would be nothing worth salvaging, because Parry Thomas's car had lain under the sand for more than 40 years and surely the sand and sea water would have taken heavy toll? I passed on this view to F.M. Wilcock who had previously contemplated digging for the remains. What I did not appreciate was that after 'Babs' had been hastily interred immediately after Thomas's death, with a cairn of stones above the 'grave', the car had been moved further up the beach, out of reach of the waves. When the military firing-range was built the concrete apron covered the area, but when this was broken up for some reconstruction work, the car was exposed. In fact, the Commandant took me to look into this hole, and very evidently, there were the remains of 'Babs'.

After Owen Wyn Owen had dug up 'Babs' and taken the car on a trailer behind a Land Rover to his garage in Capel Curig, what was left was found to be unexpectedly intact. As an engineering lecturer he knew what was the magnitude of the task ahead of him in making 'Babs' run again, and many companies in the motor trade came up with generous help. Work started in May 1969, with the chassis being completely dismantled, straightened, scraped, cleaned and painted. The engine was given similar treatment and assembled on the chassis with spare cylinder parts and as many original items as possible. A new support casting for the clutch bearing was required and the brakes were relined. The 90 mm offside hub bearing was sought and Zenith put in hand work on remaking the four carburettors, as the aluminium choke-tubes in the bronze carburettor bodies had corroded away. The coil-ignition system was sorted out, using a Lucas distributor as a substitute for the missing AC Delco one.

By December 1975 the old car and its restorer were ready to be interviewed for BBC TV by Wynford Vaughan-Thomas, and clearly Parry Thomas's old L.S.R. car was in good functional order again. Early in 1977 Pendine Community Council persuaded Wyn Owen to bring 'Babs' back there and run it briefly on the famous sands. At the time there was some thought of displaying it in a sort of memorial hall nearby, but this has not happened. I went down to see this and was able to photograph the car

The wreck had been buried by Thomas's staff after the accident, but Owen Wyn Owen disinterred the car many years later and 'Babs' was gradually restored to running order. Wyn Owen is seen here driving on to the Pendine beach, just as Thomas had done for his last time in 1927.

The author about to be taken for a few fast laps of the Silverstone Club circuit by Roger Collings, when 'Babs' was entered for a V.S.C.C. race meeting in 1988. Owen Wyn Owen, who restored this famous motor car, is on the right of the picture.

for *Motor Sport* in practically the same place on the ramp leading to the beach as 'Babs' had occupied before the fatal record bid in 1927. An American motor paper had sent a representative to Pendine for the occasion and I am afraid I rather beat him to the story.

The original Liberty engine had been replaced by a Lincoln Cars-built Liberty power unit, shipped to Wyn Owen on the QE2, and the twelve separate cylinders from it were mounted on the Packard-Liberty crankcase. Two AC Delco distributors were remounted and the original radiator was in use, with a new core which Delaney-Galley had made for it. The original oil and water-temperature gauges were on the instrument panel. Before the engine was started some 14.5-gallons of water had to be poured into the scuttle header tank, and bled through plugs in the top water-galleries.

It was splendid to see 'Babs' in thunderous action, and later it was driven at a Brooklands Society Reunion, returning to the Track it first knew, when Wyn Owen very kindly let me ride beside him along the post-1939 Runway. 'Babs' has made many reappearances since, going to Pendine again, and to other Brooklands Reunions, when it has run on what is left of the Byfleet banking (in the 'wrong' direction) with Wyn Owen's second wife as 'riding mechanic'. Later a new body was constructed, smaller drive sprockets were made to ease the load on the Thomas multi-plate clutch, which so quickly overheats, and the Budenburg Gauge Co. in Anglesey restored the dashboard instruments.

In 1980 another TV documentary was done on 'Babs', Raymond Baxter the commentator, and the venue a N. Wales beach for convenience; not Pendine. At this time the old, battered tail was still in use on the chassis but a new radiator cowl had been made. Although a key had sheared in the nearside driving-sprocket shaft all was well and Wyn Owen drove at some 60 m.p.h. in bottom gear for the sake of the cameras. In 1981 Wyn Owen took 'Babs' to the Fairbourne narrow-gauge railway, and drove Count Louis Zborowski's locomotive 'Count Louis'. Then in 1988, Roger Collings, the V.S.C.C. President, persuaded Wyn Owen to let him enter 'Babs' for races at the V.S.C.C. Silverstone Meeting. On the previous day Roger kindly gave me two interesting laps in the car in practice, but most unfortunately, on race day, after further practice lappery, a con-rod broke while the Land Rover was tow-starting 'Babs' prior to its first race.

However, I am glad to say that eventually the damaged engine was restored and Wyn Owen demonstrated the famous car at the Isle-of-Man speed events in 1990. Pendine Council had taken legal action to claim complete ownership of 'Babs' but Wyn Owen was able to negotiate for a trust to be formed, of himself, Pendine Council, and the National Museum of Wales, to administer to the future of the car and its locations.

12. The Sunbeam-Napier

Cyril C. Bone came to the aero-engined concept of a Brooklands racing car rather late, most of them having been conceived in the early 1920s. However, he displayed as much enthusiasm as any of their constructors but, alas, his efforts were to end in tragedy.

He found a big Napier in London in 1925, bought it, and apparently drove it home to Sussex where he lived, at night, and without licence or number-plates, as was possible in those faraway carefree days. He then licensed the old car as a hackney carriage, to reduce the high cost of taxing its big engine. Because, what Mr Bone had found was a pre-war M-type Napier into which someone had installed a V12-cylinder, 90×150 mm. (11,451 cc), 225 hp Sunbeam aero-engine, of the type used in small RNAS airships during the war, probably a 'Mohawk'. This engine rotated anticlockwise, making cranking-up difficult. The car had a four-seater body and was used on the road for a time, when its excellent acceleration was put to good use, although the police were not always amused, I was told.

In the autumn of 1925 Bone decided to prepare the Sunbeam-Napier for racing at Brooklands. With his friend G.B. Crow, the chassis was stripped down fully for cleaning and replacing worn parts. However, little of the original chassis was altered. A small two-seater body was constructed for it, painted white and red, with black wire wheels. The Sunbeam-Napier was ready for the 1926 B.A.R.C. Summer meeting. So as not to give too much away to the handicappers, Bone and Crow eased up behind the aeroplane sheds, to lap at around 105 to 110 m.p.h., although Mr Crow recalls that they reached about 125 m.p.h. down the Railway straight. 'Wind protection', he told me, 'was negligible and it was certainly soul stirring coming off the Home banking. . . .'

For the racing on 3 July 1926 Bone had optimistically entered for the Lightning Short Handicap and had been treated not so harshly by the handicappers as was sometimes the case; he was given a start of three seconds from Jack Barclay's TT Vauxhall and 13 seconds advantage over the two scratch cars, John Cobb in R.B. Howey's Ballot and Capt. Howey's Leyland Thomas driven by his brother. In fact, the newcomer was on the same mark as the other 'limit' cars, George Duller's 2-litre Bugatti, and Capt. Alastair Miller's 4.9-litre Sunbeam which was destined to win the race, from Barclay. Unfortunately, in the excitement of his first race Bone let in his clutch too quickly and the torque of the aero-engine twisted the propeller-shaft universal joint into knots. That was that, and the old car had to be towed away.

Undaunted, Bone entered for the August Bank Holiday meeting of

Cyril Bone's ill-fated Sunbeam-Napier. This is thought to be the only photograph ever taken of this car.

1926. Again he nominated the car for the Lightning Short Handicap. This time the handicappers were a shade more cautious, placing the 11.5-litre giant at the 31 seconds mark with Barclay's TT Vauxhall, Capt. Woolf Barnato's 2-litre Bugatti with jockey Duller 'up', and Miller's 4.9-litre Sunbeam, the eventual winner again—although the Sunbeam was, on the day, rehandicapped two seconds. Alas, when practising on the Saturday before the Bank Holiday, the car was involved in a very sad accident. It ran off the Byfleet banking, ending up in the ditch, killing the girl passenger, whose mother had expressed alarm at her wanting to ride in the Sunbeam-Napier with him.

It seems that in their innocence the car's crew had failed to appreciate the stresses involved when a large engine, even if a comparatively light aero-engine, is placed in an old chassis intended for a smaller power-unit. In fact, the front axle came adrift because a U-bolt, attaching it to a spring, contained an undetected internal fault. The front end of the chassis was badly damaged and distorted in the crash. After this tragedy the wreck was towed to a nearby Chertsey garage and there Mr Crow examined it and recovered the offending U-bolt, which was produced at the Inquest.

Had this accident not happened it would have been interesting to have seen how Bone would have fared; he had this time also entered for the Lightning Long Handicap, in which race he would have been flagged-off from the 'limit' position, in the company of not only the Vauxhall and Bugatti but with Kaye Don in the Wolseley-Viper, its vee-eight-cylinder aero-engine only slightly smaller. Alas, it was not to be and, after languishing in the aforesaid garage for a time, the car disappeared.

Its creator may not have had motor-racing experience but I gather that during the General Strike earlier in 1926, when any car, whether licensed or not, was welcome to serve as Government transport, the Sunbeam-Napier was used on the newspaper run between Fleet Street to the wholesale newsagents in Brighton, when Bone is said to have done the journey in 1 hour 15 minutes, or slightly quicker than the time taken by train, in spite of considerable late-afternoon traffic dealing with the strike. Some years later he apparently bought a straight-eight Bugatti, the ex-Jack Peacock car, which he drove from Hyde Park Corner to Preston Crescent, Brighton, in 59 minutes, at night. C.C. Bone was killed during the war, in a motor-cycle accident.

At the Inquest in 1926 Bone, a member of the Stock Exchange, after tea at the Track, asked Mr Armitage, whom he knew at Cambridge, and his friend Miss Norris, aged 20, daughter of the late Surg. Cdr. Norris, if they would like to go round in his racing car. As the car was a two-seater, only Miss Norris went—he did not know that her mother did not want her to. They had gone one lap on the Byfleet banking when on the next lap, doing just over 100 m.p.h., preparatory to coming in, Bone felt the front of the car swinging down the banking. He tried to correct the apparent skid but the steering did not respond to full right lock.

A Mr D. O'Donovan (the racing motorcyclist perhaps?) said he jumped out of the way of the car, which swept away the railings and disappeared into a culvert. Bone's clothes were in flames, the car having overturned and caught fire. With a Mr Bruce he put out the flames. Miss Norris had been flung out and killed instantly. It was concluded that a U-bolt had broken and released the front axle. Bone was in hospital in Weybridge for some time with a fractured collar bone and burns.

Mr Albert V. Ebblewhite, the Brooklands' handicapper, said he had timed the car to lap at 100.61 m.p.h. before the accident. It held the Track well. He observed, though, that 'a chassis of one make and an engine of another was always an unsatisfactory combination'! Bone said he had got 113.5 m.p.h. from the car in earlier practice. The Coroner gave the verdict as Accidental Death but commented that it had been imprudent to put this engine into a chassis of 1909/10 and to run at such high speed.

13. The Napier-Railton

The Napier Railton was built for the late John Cobb, holder of the Land Speed Record and the Brooklands outer-circuit lap-record, in 1933, primarily with a view to breaking the World's 24-hour record.

John Cobb had always been fascinated by the thought of racing really big cars round Brooklands track. He won his first race there at the wheel of a 1910 10-litre Fiat, and subsequently had a very notable Brooklands career with this car and the 10.5-litre V12 Delage. Cobb also drove Austro-Daimler, T.T. Vauxhall, Thomas Special 'Babs', 4-litre Sunbeam, Riley 9, 'Monza' Alfa Romeo, various Talbots, 1.5-litre GP Delage, 4.5-litre Lagonda, and possibly other cars at Brooklands and at other important venues. But this chapter is concerned with one of Cobb's most consistently successful cars, and one which established records at Brooklands, Montlhéry, and Utah Salt Flats—the Napier-Railton.

The old 10.5-litre Delage was getting rather long in the tooth and was certainly not the car for attacking records which involved high speed for two rounds of the clock, so Cobb sold it to the young barrister, Oliver Bertram, and contracted with Thomson and Taylor of Brooklands to build him a new track car that would combine the engine reliability of the old aero-engined giants with chassis stamina such that fast lappery on the rough surface of Brooklands and Montlhéry would not result in breakage and disaster.

Reid Railton was put in charge of the design and Ken Taylor supervised the constructional work, many of the components being machined from the solid, and beautifully finished in Thomson and Taylor's workshops.

Long before the big car was complete a scale model had been made of it, complete with realistic wire wheels and dummy engine, by an Italian model maker. Later Cyril Posthumus completed this model for Cobb by putting a body on it.

To revert to the real car, this was highly ingenious and quite remarkable for the year 1933. Railton decided to build it round a Napier 'Lion' aero-engine, for these powerful and reliable engines, dating from circa 1919, were well known to Thomson and Taylor Ltd, who had converted many of them for marine purposes, and these engines and parts were still fairly easily obtainable. Thus another make of aeroplane engine was added to those of Sunbeam, Hispano-Suiza, Wolseley, Mercedes, Fiat, Benz, Clerget, Liberty, and Maybach that had seen high-speed service in racing cars on Brooklands track.

Originally Railton had visualized the older 450 hp Napier 'Lion' power unit, but he was able to use the version that developed 502 b.h.p. at 2,200 r.p.m. In fact, on test at Acton before T & T's installed it in the chassis, the 'Lion' gave 564 b.h.p. at 2,350 r.p.m. on a compression ratio of 6 to 1, and a maximum of 590 b.h.p. at 2,700 r.p.m. This was a useful advance on the 400 hp developed by the original Napier 'Lion' in 1919 and the 450 b.h.p. that was obtained by 1921, and the 470 b.h.p. of the engines used in the 1922 Schneider Trophy seaplane race. As Cobb's aim was reliability, Railton was not tempted to try to obtain a 1927 'Lion' which developed 875 b.h.p. at 3,300 r.p.m. on a 10 to 1 compression ratio, or a 1929 Schneider Trophy version as used in Segrave's L.S.R. 'Golden Arrow' that would have given 1,295 b.h.p. at that speed, or the eventual supercharged Napier 'Lion' engine capable of producing an impressive 1,450 b.h.p. This big 139.7×130.2 mm. (23,970 cc) 12-cylinder engine, its three banks of four cylinders in the form of a 'W' with the outer banks at 60 deg. to the centre bank, Railton installed in a massive chassis possessing deep side-members. John Thompson supplied these side-members, which were united by a slightly curved channel-section cross-member at the rear and by means of a tubular cross-bar at the front. It became fashionable to undersling the chassis frame of a GP car in 1926—Talbot, Delage, Thomas-Special, Eldridge Special—and now Railton used this method for his big track car, the side-members tapering at their front and back ends, also curving upwards somewhat at the front, so that both axles were above them. Two massive tubular cross-members and a channel-section cross-member completed the chassis frame.

A straight front axle, assembled in three pieces, with reversed Elliott steering stub-axles, was sprung on half-elliptic leaf springs, shackled at the front. At the back twin Woodhead cantilever springs on each side carried the back axle, which was of very special construction, being of fully-floating type built up from three high-tensile steel forgings, with an electron sump to provide plenty of lubricant. The gears were made by E.N.V. to give a final drive ratio of 1.66 to 1, and there was a differential. This axle alone was said to have cost £400. At the front, one pair of Hartford shock-absorbers ahead of the front axle, another pair behind it, were used, while at the back a pair of Hartfords set transversely and another pair mounted normally looked after damping of back-axle movements, Cobb being provided with cockpit telecontrol of one pair of shock absorbers on each axle. Short radius arms located the front axle.

There was much beautiful machine-shop work about various parts of

The Napier-Railton, designed by Reid Railton and built by Thomson & Taylor's at Brooklands, which was a highly successful racing and record-breaking car in the hands of its owner, John Cobb, who in 1935 set the all time lap record at the Track of 143.44 m.p.h.

A W-pattern twelve-cylinder Napier 'Lion' engine of the kind which took the Napier-Railton to so many victories.

the chassis, which is still very nice to look upon. The steering column was stayed to the dashboard. The rev-counter, or tachometer, read from the top right-hand segment of the dial, with clearly spaced-out calibrations of 2, 5, 10, 15, 20, 25 and 30, i.e. indicating speeds of from 200 r.p.m. to 3,000 r.p.m. although I think the engine normally ran at around 2,000 to 2,200 r.p.m. To the left of the tachometer there was a small oil-temperature gauge, marked in steps of ten degrees, from 30 to 100, the more critical readings (80/90/100 deg.) being more widely-spaced than the other figures. Under this was a big domestic brass tumbler-switch for the ignition, marked 'up for off'. Three small matching dials, set to the right of the main dial, indicated fuel-pressure, oil-pressure and water-temperature. The oil-gauge was marked from ten to 100 lb./sq. in. in steps of ten pounds, the water thermometer with similar spacings, indicating from 20 to 100 deg. C. Again, the vital temperature rises of 80/90/100 deg. were more spaced out than the others. To the right again of the water thermometer was the big knob of the Ki-gass (a means of injecting petrol into the engine to assist starting), and on the cockpit floor was a large tap, conveniently placed for the driver to find with his right hand, to cut off the fuel flow. Low down on the left-hand side of the cockpit was a 160 m.p.h. speedometer.

Front brakes were deemed unnecessary, but the pedal operated expanding rear-wheel brakes, and for holding the car when it was stationary there was a small expanding brake behind the gearbox, controlled by a tiny brake lever with a conventional ratchet. Fuel requirements were governed by anticipated tyre life, so a 65-gallon tank in the tail sufficed. Oil was carried in a 15-gallon tank between the propeller shaft and the nearside chassis member.

The Napier-Railton ran on Rudge centre-lock wire wheels shod with specially-made Dunlop treadless track tyres, 35 in. × 6 in. on the back wheels, smaller tyres on the front wheels.

To revert to the Napier 'Lion' engine, this had two o.h. camshafts above each of the cylinder blocks, ignition was by dual magnetos mounted transversely at the front, and the Napier carburettors were also at the front of the engine, feeding the cylinder blocks through long induction pipes, the inlet ports of the outer banks of cylinders being on the inside of the blocks.

Railton's problem was to adapt this aeroplane engine for car propulsion, and this he contrived to do by mounting an open Borg and Beck single-plate clutch on the rear of the crankshaft, the drive going to a separate Moss three-speed gearbox, from whence a Hardy-Spicer propeller shaft took the drive to the back axle. The gearbox was very compact, because Railton realized that to cope with the tasks for which the 24-litre Napier-Railton was intended unusually high tooth-pressures were permissible on the gears, because the indirect gears would be needed only to get the car rolling.

Piping was Petroflex and the facia carried a Jaeger rev-counter and Smith's instruments. To cool the big engine a large, slightly-V, radiator set at a very slight angle and partially cowled, was used. The overflow pipe led to a separate tank connected to the suction-side of the water pump, to conserve water during long runs. To enable radio 'intercom' to be used the K.L.G. sparking plugs were properly suppressed, and a Rotax lighting plant was incorporated, with all-night record attacks in view. This exciting chassis had a wheelbase of 10 ft. 10 in., a track of 5 ft., and it went to Gurney Nutting to be fitted with a fairly conventional, single-seater racing body. The complete car was 15 ft. 6 in. long, 7 ft. wide and weighed approximately 30 cwt. At first the great car was unpainted, but later it was given a silver-grey finish; the wheels were black. The bucket-type driver's seat had a central slide to enable it to be adjusted to suit different drivers, a big key engaging holes in the floor locking it. The accelerator pedal was central.

Naturally the news that John Cobb, experienced outer-circuit exponent, was to race a specially built aero-engined giant racing car built actually within the Brooklands Track aroused the greatest interest. I recall weighing up its prospects of being able to raise the lap record in *Brooklands—Track and Air*, of which paper I was at the time the Acting Unpaid Assistant Editor. I thought it had a good chance of beating the record, then held by Sir Henry Birkin Bt. in Dorothy Paget's 'blower' 4.5-litre single-seater Bentley at a speed of 137.98 m.p.h.

The completion of the Napier-Railton took longer than was anticipated, and not only was it unready for the 1933 B.R.D.C. British Empire Trophy race but was not, as had been hoped it might be, on view in the Paddock.

However, Cobb was able to bring his new car out for the 1933 B.A.R.C. August meeting. It was described in the Race Card as 'Aluminium; Black'. Incidentally, if the subsequent Land Speed Record Railton Mobil was built for £10,000 I imagine the Napier-Railton had not cost more than about £5,000. The Napier 'Lion' engine ran on normal fuel and func-

A pit-stop at Brooklands for the Napier-Railton, during a B.R.D.C. 500-Mile race.

tioned reliably for many years, but tyres were a very sticky problem.

Cobb was naturally on scratch in his first race in the car, the Byfleet Lightning Long Handicap, giving 3 seconds start to Kaye Don's 4.9 litre Bugatti. This was no deterrent, for Cobb set up a new standing-start lap record of 1 min. 22.6 sec. (120.59 m.p.h.), did his flying lap at 123.28 m.p.h. and won by 2.6 sec. from Don, and Bertram in the ex-Cobb Delage. A good start!

Indeed, Cobb's race average was 122.3 m.p.h. and he is said to have had to brake as he came up with slower cars. How many present-day racing drivers, I wonder, would relish driving this monster round Brooklands on the tyres of 50 years ago and with little hope of stopping if a smaller car strayed up the steep bankings into its path?

So astonished were the handicappers, Mr T.D. Dutton and Mr A.V. Ebblewhite, that Cobb was re-handicapped from scratch to 'owes 14 seconds' in his next race, whereas Kaye Don and Bertram were put back only two seconds. This was too much for Cobb, but he nevertheless did his standing-start lap at 117.19 m.p.h., his next circuit at 137.20 m.p.h., which was a new Class A lap record, and, unplaced, ran in at a lap speed of 135.70 m.p.h.

The Napier-Railton did not race at Brooklands again that year. This was because its owner was anxious to employ it for long-distance record attacks. Cobb had the World 24-hour record in view, and as Brooklands

Cobb trying out the improvised gear, fitted in order to facilitate pit visits in the course of long-distance record bids.

wouldn't permit night running his big car was shipped to Montlhéry track, near Paris. The Napier-Railton had undergone tyre tests at Brooklands beforehand and as the complex system of Brooklands' silencers would not be needed in France, shields had to be fitted to protect Cobb's eyes from dazzle set up by flames from the stub exhaust pipes, while another, stronger shield was placed so that the driver's arm would be protected if a back tyre flung a tread. Very big wheels were necessary, which were difficult to remove from the hubs during pit-stops.

The co-drivers were Cyril Paul, the Hon. Brian Lewis (Lord Essendon), and Tim Rose-Richards; and Dr Whitehurst went as M.O., with Jean Chassagne acting as liaison officer. George Eyston was persuaded to lend his trackside lighting equipment. There was the problem of stopping the big car in the correct place for refuelling with benzole, its weak brakes not helping, and another difficulty was restarting the engine, because a push-start was necessary, but as the entire run had to be done under power this involved running the Napier-Railton back a considerable distance before a push-start could be attempted.

But as S.C.H. Davis made clear in his book *The John Cobb Story* (Foulis), the crux of the matter was tyre-life. If the front tyres lasted 400 miles, the back tyres 200 miles, the speed schedule to beat the record would be 126 m.p.h. for two rounds of the clock, allowing for 15 depot stops. But if the back tyres lasted only 160 miles, 19 stops would be necessary, and thus a higher speed would have to be maintained—yet every increase in speed meant more tyre wear, and therefore more stops, calling for further speed increases. A vicious circle indeed.

However, a start was made, the big car lapping steadily at 130 m.p.h., and in 96 minutes the first record fell—the World 200-mile record at 126.44 m.p.h. The tyre change occupied a mere 52 seconds and Cyril Paul replaced Cobb as driver. Soon afterwards a rear tyre burst, but two fresh wheels went on in 32 seconds. Thereafter tyres gave up with depressing frequency. The three-hour record fell but speed had to be reduced, so that for 500 miles was missed. Then the track surface began to break up, resulting in a shocking ride for whoever was driving. The headlamps on their detachable bar were fitted, the trackside lamps laid out, but in the end it was deemed wise to abandon the attempt. World records up to 1,000 km. and six hours and many Class records were broken, and had it been known that Paris had refused to recognize speeds claimed by A.C. Jenkins's Pierce-Arrow in America, a lower speed schedule would have been set, and the 24-hour run might have been possible.

Back in England Cobb had the Brooklands' silencers refitted and in October set about attacking the standing start kilo and mile records. The first onslaught was a failure, but smaller tyres lowered the gear ratios and gave Cobb the World and Class A standing-start mile records at 102.52 m.p.h. The kilometre record was thought to have been captured also, but actually a faster time had been done by a Maserati at Montlhéry. This spurred Cobb on to come out again in November, when, using the large-size tyres, the Napier-Railton set the record to 88.52 m.p.h. Coming off the banking in the reverse direction was extremely difficult and reduced the speed for this two-way run, but over the normal flying kilometre Cobb was timed at 143.67 m.p.h., the highest officially-clocked speed at Brooklands up to this time.

During the winter, before John Cobb's 34th birthday, different gears were installed and Dunlop made bigger tyres for the car.

Tests of the new tyres having been satisfactorily completed during a run of 75 miles, during which Cobb lapped at 134 m.p.h., and he, Cyril Paul and the Hon. Brian Lewis averaged about 130 m.p.h., the attack on the lap record was scheduled for the B.A.R.C. Easter meeting.

Cobb was billed to make his run at 3.10 p.m., electrically timed. He succeeded, with a lap at 139.71 m.p.h., compared with Birkin's 1932 lap at 137.96 m.p.h. in the Bentley. I had predicted previously in *Brooklands—Track and Air* that, given a cool day and ordinarily favourable conditions, the Napier-Railton should be able to lap at 138 to 139 m.p.h. and that whether or not it would go faster was a matter of how it took, and withdrew from, the Brooklands bankings. Incidentally, the big car was now finished in green mottled aluminium with black wheels and had won the prize for the smartest racing car present.

On its second lap a gusty wind had caused Cobb some anxiety as he came over the Fork, but down the Railway Straight he attained a speed close to 160 m.p.h. His arms and fingers were stiff from fatigue at the end of these few laps.

Nevertheless, Cobb came out again, to start from scratch in the Ripley Lightning Long Handicap. He opened at 113.19 m.p.h., did the following lap at 133.52 m.p.h., then, out-handicapped, eased up to 129.70 m.p.h.

John Cobb, owner and intrepid driver of the mighty Napier-Railton, and Brooklands lap record holder for all time.

After this, with radiator strengthened and even larger wheels and tyres, the Napier-Railton was taken again to Montlhéry for another onslaught on the World 24-hour record. Cobb was on this occasion accompanied by Charles Brackenbury, Freddy Dixon, and Cyril Paul, and that run ended in Dixon crashing, but not before five World records from 1,000 kilo to 12 hours had been broken at average speeds of from 120.71 to 120.01 m.p.h.

The damage to the car was repaired in time for the B.A.R.C. August meeting, and Cobb drove in the Brooklands Championship race over four laps of the outer-circuit, winning by 11 seconds from Dudley Froy in Kaye Don's 4.9-litre supercharged Bugatti, after lapping at 118.30, 140.93, 138.34 and 134.60 m.p.h. Thus not only was the lap record raised again, but Cobb put up a new highest race speed of 131.53 m.p.h.

All wheels clear of the concrete—the 24-litre Napier-Railton in fast action at Brooklands Track.

The Napier-Railton ran again in the Esher Lightning Long Handicap, lapping at 117.19, 134.60 and 134.24 m.p.h., but was unplaced.

Now John Cobb set himself another very strenuous and hazardous task, that of driving the big car in the B.R.D.C. 500-Mile Race. Record attacks on an empty track were one thing but now came the hazard of having to overtake continually all the slower entries. Tim Rose-Richards shared the task. Three tyre changes would be required to the single stop made by most of the field. Cobb set off, lapping at 116 m.p.h., but rain began to fall, turned to a downpour, and under the circumstances the Napier-Railton was wisely withdrawn.

Cobb did not enter for the B.A.R.C. Autumn meeting, being content to rest the Napier-Railton until its lap record was broken—and it remained intact in spite of Whitney Straight's attempt with the Duesenberg at this meeting.

In 1935 Cobb decided on a very ambitious step—that of having his third crack at the World 24-hour record, this time not at Montlhéry but on the faraway Salt Flats at Utah, USA.

Taking as his co-drivers Rose-Richards and Charlie Dodson, Cobb contrived to lap the salt circuit at 150 m.p.h., so that in spite of drastic tyre trouble the Napier-Railton took the World 24-hour record at last, at an average speed of 137.40 m.p.h., taking twenty other World records on the way, including the hour record at no less than 152.7 m.p.h. Indeed,

The 'Right Crowd' admiring the Napier-Railton.

for the first 50 kilometres the car had averaged 154.46 m.p.h. Returning home, Cobb drove the Napier-Railton in the Brighton Speed Trials. Not the fastest car present, it was second in the Unlimited Racing Car Class with a speed for the standing start half-mile of 76.27 m.p.h. Incidentally, the Napier-Railton, lighting-set still in place, also did some demonstration laps at the JCC Donington Park race meeting late in 1935 after its return from Utah.

The car was back at Brooklands for the B.R.D.C. 500-Mile race. I remember the fantastic sight of the huge aero-engined car, now in silver aluminium, coming round lap after relentless lap at the very top of the bankings, overtaking lesser cars as if they were travelling backwards. Cobb held second place to Oliver Bertram's Barnato-Hassan, which had wrested the lap-record from it earlier in the year and now had a 4-second start. Then Cobb took the lead, averaging 126.89 m.p.h. But its first pit stop, for four fresh wheels, staggeringly heavy to lift, and 50 gallons of fuel, cost ten minutes, and allowed Bertram to regain his lead. Soon, however, the fuel tank of the Barnato-Hassan gave way under the stress of the Brooklands' bumps and the car was retired. Cobb was able to ease up slightly and when a piece of flying concrete cut his face he gave over to Rose-Richards again, the enormous wheels being changed in a mere $1\frac{1}{2}$ minutes. So the Napier-Railton, running on its home ground, won the '500', at a speed of 121.28 m.p.h., after a race lasting 4 hr. 28 min. 52

John Cobb and stockbroker T.E. Rose-Richards after they had taken the Napier-Railton to victory in the 1935 B.R.D.C. 500 Mile race at 121.28 m.p.h.

sec. Not until 1949 was this average bettered in the Indianapolis 500-Mile race, and then by only 0.05 m.p.h., and from a rolling start!

The victorious car appeared again at the B.A.R.C. Autumn meeting but a gusty wind hampered Cobb's efforts. In the Second October Long Handicap he started from scratch, lapped at 114.23, 133.52 and 132.80 m.p.h., but was unplaced.

Bertram now held the lap record, at 142.60 m.p.h. Cobb, therefore, took to the track on 7 October and, although the surface was not quite dry, he lapped at 143.44 m.p.h., the Napier-Railton sliding quite a lot and disposing of its tyres in a couple of laps—showing how close was the margin of disaster to safety in this department if speed was lifted above the 130 mark. This lap record will now stand for all time, and in establishing it Cobb was timed over a kilometre at 151.97 m.p.h., the fastest speed ever clocked officially on Brooklands.

But the full speed-potential of the Napier-Railton had not yet been realized, as Cobb was able to demonstrate in September 1936, when he made another expedition to Utah and, partnered by Rose-Richards, Hindmarsh and Charles Brackenbury, regained the World 24-hour record at 150.6 m.p.h. capturing seven other World records including the

Barrister Oliver Bertram congratulates John Cobb after they had won the 1937 B.R.D.C. 500-Kilometre race in the Napier-Railton at 127.05 m.p.h.

'hour', on a separate run, at no less than 167.69 m.p.h., the best speed being 168.59 m.p.h. for 100 miles. On this occasion the difficulty of starting the engine away from the depot was solved by rigging up a starter motor that the driver could engage, via a roller, to turn the offside back wheel—see page 117.

The B.R.D.C. '500'—changed in 1937 to 500 kilometres to reduce the cost of competing and the savage wear and tear on the cars—was the only time after this that the Napier-Railton competed at Brooklands. The 1936 race was missed because Cobb was out at Utah. This time very careful plans were laid to ensure that tyre trouble should not spoil the car's chances of repeating its 1935 victory. Ken Thomson controlled speed very carefully with lap by lap signals, and the tyres were carefully watched. Cobb chose Bertram as his co-driver and, although a tricky wind made the car difficult to hold, Bertram lapped at 126 m.p.h. Even though a tyre failed before a scheduled pit stop, between them the drivers won convincingly, at 127.05 m.p.h.; Cobb doing the last lap at 135.45 m.p.h. with the tread of the offside front wheel flapping as he crossed the line. I can recall seeing the great car being towed in with a happy Ken Taylor in the cockpit, while Cobb and

Bertram received their plaques from Lord Howe in the Finishing Straight—a stirring race and a fitting finish to the Brooklands career of the Napier-Railton.

John Cobb continued his own speed career, taking the Land Speed Record with the remarkable Railton Mobil Special, a four-wheel-drive car using two supercharged Napier 'Lion' engines, which now resides in Birmingham Transport Museum. Indeed, to this day, Cobb, who lost his life attacking the Water Speed Record on Loch Ness, is holder of the Brooklands lap-record. Also his Napier-Railton exploits won him the B.R.D.C. Track Gold Star in 1935 and 1937.

His cars continued to be cared for by Thomson and Taylor Ltd when war broke out, and in 1942 Granville Grenfell kindly took me to a hide-out not far from Brooklands where the Napier-Railton was stored, along with other racing cars. I remember that its wheels were shod with 7.50×20 in. tyres. The car, then nearly ten years old, had had a very

After Brooklands had closed because of the war, the Napier-Railton found a new lease of life testing GQ aircraft parachutes. Here is the rig.

The ex-racing car getting on with its new task. The driver is probably Geoffrey Quilter of GQ Parachutes.

creditable career, during which its engine functioned with practically no trouble and a minimum of attention—unique in motor racing circles! Its appearance was also very little changed, although its original radiator cowl, barred on some occasions, was later opened out somewhat.

After the war there was nowhere much for the Napier-Railton to race, with Brooklands having failed to survive the conflict and Cobb being engaged at Utah with the twin-engined Railton Mobil. But it came in useful, on the advice of S.C.H. Davis, for a film called *Pandora and the Flying Dutchman*, after it had been dolled up as a fictional Land Speed Record car with a new nose cowl and longer tail. Shots were filmed at Pendine Sands and in Spain, the stub exhausts adding realism and the car surviving being driven into the sea while loaded with fireworks to simulate a sensational conflagration.

Later the car was transferred, through Thomson and Taylor Ltd, to the G.Q. Parachute Company, who needed a very fast, reliable car with which to test their aircraft tail 'chutes. A structure was rigged over the tail from which parachutes could be released from a stowage and rewound thereto by an electric motor. The original brakes were considered unsuitable for stopping the car on a short runway when destruction tests caused the brake parachute to part company, so new Dunlop disc brakes with Mintex pads were applied to the back wheels. Extra equipment was needed in the cockpit, and Dunlop 7.00 in. × 19 in. racing tyres were fitted. The old 'Lion' engine proved as reliable as ever for these sprints of up to 140 m.p.h., apart from water leaks. It was run on pump grade-one Shell petrol, Castrol XX oil and Champion R7 plugs. Sir Raymond Quilter Bt., of G.Q., thoroughly enjoyed driving the car and demonstrated it to the Press at Dunsfold in May 1954.

The next owner of the famous Napier-Railton was the Hon. Patrick Lindsay, here seen racing it at a V.S.C.C. Oulton Park meeting.

He also drove the Napier-Railton at the Vickers-Armstrong Brooklands Jubilee Celebrations in July 1957, towing it there and starting it with a Ford Zephyr, but after his death it returned to the care of Thomson and Taylor Ltd, alas no longer situated within the Track.

After G.Q. Parachutes had finished with the car it was acquired by that great enthusiast the Hon Patrick Lindsay. He found the cylinder blocks had become porous, so he had Douglas Hull look over the car. Test runs were done at Blackbush aerodrome, and Lindsay then bravely ran the Napier-Railton at a V.S.C.C. race at Oulton Park in 1962, contriving to lead a race until he spun off at the last corner. I recall the splendid sight of the big car emitting smoke, water vapour, and flames from the stub exhausts, with the disc brakes almost on fire towards the end. The smaller wheels and tyres used by G.Q. and the long tail used for the Pandora film were retained at this time. Lindsay ran the engine on Castrol GP oil and 7.00×19 front tyres.

Lindsay had also raced the Napier-Railton on the Silverstone outer-circuit at an old-car event, before which he kindly allowed me to sample it, after it had been towed to the course from Lord Hesketh's nearby estate (*Motor Sport*, June 1970, page 801).

He ran it again at the 1960 Aston Martin Owners' Club Martini Trophy Race meeting, finishing third behind two ERAs. Around this time the car was housed in the Racing Car Hall at the Montagu Motor Museum. Pat Lindsay intended to overhaul the aged engine, and diabolical jigs had to be made to uncrust the valve-cages from the cylinder blocks. After Lindsay's sad death the next owner was Bob Roberts, a very enthusiastic collector of good motor cars, who kept many of them in his Midlands Motor Museum at Bridgnorth. To bring the car back to his exacting standards he had it overhauled by the Hodec Engineering Company of Old Woking, much of the work being done by Peter Horne, who had

Lindsay let the author unleash the car at Silverstone. Getting briefed . . .

worked at T & T's up to 1941.

The complicated exhaust system, which had been quite a problem when the car was being built originally, had to be completely refabricated, for the three banks of four cylinders. The original brake drums were missing, so the disc brakes which Ian Taylor, Ken Taylor's son, had helped to install for G.Q., were retained. The parachute testing was apparently done mostly in second gear of the three-speed Moss gearbox,

. . . and on his way.

Two last impressions of the great car, showing its complicated exhaust systems, necessitated by three cylinder banks of the Napier 'Lion' engine. The person standing behind the car is Hugh McConnell, the Brooklands Scrutineer. For many years the Napier-Railton was owned by the late Bob Roberts, and on view in his Midlands Motor Museum. It was then bought by Victor Gauntlett, and sold at auction.

as it was badly worn and had to be replaced. The water leaks were attended to, after which the jackets were pressure tested, but otherwise very little had to be done to the engine, although the water-pump drive had sheared. This is remarkable, because it seems quite definite that the original Type E89, Series-XIA Napier 'Lion' engine, installed in 1933, was the only one used in the car. Indeed, it had presumably never previously been taken down, although at the most modest estimate it must have run at least 12,000 racing-miles, at around 2,000 r.p.m. The B.T.H. magnetos had at some time been replaced by Watfords IB12 magnetos.

The Borg & Berk clutch was in sound order but the wheel bearings and some of the gearbox bearing were replaced. Mr Horne told me that the repairs to the chassis after Freddie Dixon's crash at Montlhéry were still visible but not the headlamp mountings. The instruments were thought to be original, but there were two blanked-off holes in the panel which may have been where G.Q. Parachutes installed extra instruments for their own use. The panel had become dishevelled, so a new one was made up. Thus the old car was returned to pristine condition. The 7.00×19 Dunlop Racing tyres were retained and the Napier-Railton became the premier exhibit at the Midland Motor Museum. Many drawings and other documents had been retained with the car, and also one of the plain-tread Dunlop 7.50×20 tyres, a reminder of the Bonneville record runs.

Some time before this book was completed Victor Gauntlett bought the Napier-Railton, following Bob Roberts' death, and I believe it was purchased, at auction, by someone living in Germany.

14. The Rest

There were other cars endowed with aero-engines that made appearances at Brooklands, but not in races; that is, if we ignore the 54.6 hp Adams which the pioneer Bleriot pilot Graham Gilmore ran in some minor events at the Track in the very early days, which may or may not have had a power-unit intended for an aeroplane. Granville Bradshaw used an airscrew-chassis for propeller-testing on the Motor Course (of engines like the 30/40 hp A.B.C., which once broke the propeller blades off) but that wasn't raced there. The remaining cars worth describing were likewise not raced at the Track but they made visits there and, indeed, may well have 'gone round' on non-race days.

HARRY HAWKER'S SUNBEAM-MERCEDES

Probably the first of such road-going aero-engined giants was the car built by the great Australian test-pilot Harry Hawker, while he was working at Brooklands for the Sopwith Aeroplane Company. He must have commenced work on the car, which he constructed himself in his workshop at his house in Hook, before, or soon after, the Armistice was signed in November 1918. It was running in chassis form by January 1919. Hawker, who had been competing in *The Daily Mail* Circuit-of-Britain aeroplane race at this time, had obtained two 225 hp side-valve V12 Sunbeam-Coatalen aeroplane engines, of the Mohawk-type as used by Coatalen in his pre-war 9-litre racing Sunbeam, one of which had flown 18 hours in a Short seaplane. It would not have been too difficult, one assumes, for a famous test-pilot to come by such engines.

Hawker then cast about for a 35 hp Mercedes chassis in sound condition, choosing this make for the known strength of its gearbox and transmission. He then lengthened this chassis and moved its gearbox back ten inches, in order to install the Sunbeam aero-engine. The rear half-elliptic springs were removed and, by the provision of additional leaves, converted into cantilever springs. The weight of the Mercedes flywheel having been halved, it was attached directly to the engine crankshaft, the aircraft reduction-gear having been discarded. For starting the big engine a large geared CAV starter, as used in war-time tanks, was fitted, engaging with teeth cut on the flywheel.

The Sunbeam engine, which had its cylinders in four banks of three, was mounted on a steel sub-frame supported on two cross-members. It pretty well filled the long bonnet, which needed two small 'helmets' on each side to clear the components. It was only with difficulty that the right-hand raked steering column was got past the rear cylinder head

The Sunbeam-Mercedes which the celebrated pilot Harry Hawker built-up before, or soon after, the 1918 Armistice. He intended to race it but decided that the 350 hp V12 Sunbeam was a better proposition. He then used the Sunbeam-Mercedes as a fast road car and it was destined to have a dramatic career, after Hawker had come down in the sea during his attempt to fly the Atlantic in 1919.

and offside induction pipe. For aircraft work these engines had two carburettors per cylinder bank, which Coatalen had retained for the racing Sunbeam, but Hawker decided on a single updraft Claudel Hobson carburettor per six cylinders, each one feeding through a four-branch inlet manifold with a central water-heated hot-spot.

Hawker built up a very neat radiator for his Sunbeam-Mercedes, its shape not unlike that of his Austro-Daimler car, with a tall filler neck and a large hole to accommodate the water pump. Equipped with a tyre-pump with which the passenger could maintain pressure in the petrol tank, two shallow bucket seats, and a few rudimentary instruments, this exciting hybrid was ready for Harry Hawker to try out.

The provision of cantilever back springs, which gave a comfortable ride, the use of two carburettors to increase gas flow and so improve the engine idle, and the early intention to have a four-seater body made for the car, suggests that Hawker never intended to race the monster, inspite of a claim to the contrary made in a recent biography of the famous airman. He did try the car out on Brooklands, and had an alarming experience when it skidded violently after a tyre burst. But it was as a fast, accelerative touring-car that the Sunbeam-Mercedes seems to have excelled. On early tests in and out of the Kingston labyrinths close to the Sopwith factory, the big chassis proved easy to drive, and quiet, as many aero-engined cars were. The engine ran cool, although no fan was fitted, and a rough check showed a petrol-thirst of 16.5 m.p.g. The hill-climbing powers were apparently remarkable, but one difficulty, especially in chassis form, was poor adhesion of the back wheels.

In the course of construction there had been the inevitable problems. But in the end, after many holes had been drilled in the chassis side-members, which later had to be filled-in, the Mercedes four-speed gear-box was used instead of a smaller two-speed box. The ground clearance beneath the engine was found to be very small, only about six inches, even after it had been raised a little. Muriel Hawker, whom Harry had met during the war when her car had run out of petrol in Richmond Park, and he and a friend had returned in Hawker's Gregoire to assist her, and whom he soon married, helped her husband to rivet-up the seven feet or so of aluminium bonnet. When the time came to start the engine for the first time after it had been installed in the Mercedes chassis, Harry had to crank it as the batteries were nearly flat and the starter only just able to turn it over. It fired-up eventually but on only one bank of cylinders, and soon became so hot that the solder ran in the joints of the induction pipe before Mrs Hawker stopped it, Harry having gone off to fetch some tools. The next day the damaged joints were rebrazed and all was well. The hybrid took its proud place in the 'Ennersdale' garage beside the 25 hp Talbot saloon, a sports Model-T Ford, and the Gregoire in which the couple had spent their honeymoon, running partly on gas contained in a gas-bag, as the use of petrol was apt to be questioned.

The Sunbeam-Mercedes became a reliable car, after its twelve K.L.G. sparking-plugs had been provided with special adapters to prevent them oiling-up at low revs. By the summer of 1919 Hawker had had a four-seater body of his own design made in polished aluminium for the car by the London Improved Motor Coach Builders Ltd, of Lupus Street, in Westminster. A screen for the rear seat occupants was a feature, there were no running-boards, and helmet-type mudguards were to be fitted, those at the front being arranged to turn with the wheels, although at first curious strip ones were used, bent over to almost touch the tyres at the top and almost touching the ground at the bottom. I am certain this was a means of trying to pick nails out of the front tyres and sweep any away from the path of the back tyres, a ploy sometimes adopted at this time when horse-shoe nails were apt to cause punctures. By 1921 normal front mudguards and running-boards had been fitted.

The Sunbeam-Mercedes figured in many episodes that proved its capabilities, such as beating a Benz on the road between Putney and Kingston in an impromptu race when it had a Rolls-Royce chassis on tow, and it was the car in which Muriel Hawker rode when seeing her husband off on his ill-fated transatlantic flight, that ended in disaster.

Hawker and his navigator Comdr. McKenzie-Grieve started the crossing from Newfoundland on 18 May 1919, their destination Brooklands, but the Rolls-Royce engine in their Sopwith Atlantic biplane overheated and they had to force-land in the sea. They were picked up by a small ship without radio and were thus presumed drowned. HM King George V sent a telegram of condolence to Mrs Hawker. Then, when the news that both airmen, and the aeroplane eventually, had been rescued, HM The King sent another telegram, this time of rejoicing, and it was in the Sunbeam-Mercedes that Mrs Hawker, driven by Harry's brother-in-law,

Capt. L. Peaty (who raced a Bleriot-Whippet cyclecar at Brooklands) went to the celebrations, a big Australian flag on the radiator and a Police Inspector excusing their high speed up Putney Hill. The car was also prominent when Hawker and Grieve arrived back in London to a tumultuous welcome at King's Cross Station, where they were placed in it and when it was soon in danger of being damaged by the enormous crowds.

After Brooklands had reopened in 1920 Hawker stripped the Sunbeam-Mercedes and tried it on the Track. With the touring body it reached about 107 m.p.h. before a burst tyre on the banking caused the aforesaid alarming skids. But in Mrs Hawker's words, 'although this did not deter him, Harry replaced the touring equipment, saying that a car only capable of under 110 m.p.h. was only fit for a touring car'. Hawker was, of course, to drive for Coatalen the faster 350 hp V12 Sunbeam single-seater, as described in Chapter 2.

On 12 July 1921, Harry Hawker was tragically killed when the 360 hp A.B.C. Dragonfly-powered Nieuport 'Goshawk' biplane he was testing in readiness for the Aerial Derby caught fire over Hendon. He had ridden to the aerodrome from Hook that morning on a two-stroke Hawker motorcycle. He was buried at Hook in Surrey, and again it was the big Sunbeam driven by Capt. Peaty, the car almost hidden in floral tributes, that conveyed the chief mourners to the cemetery.

After Hawker's death the car presumably went into storage either at the Sopwith works at Kingston or at Brooklands. It was next heard of as the breakdown-vehicle at Hunt's Garage in Leominster, where it served in this capacity at least up to 1934, if not later. In very recent times it has been possible to talk with people in that town who remember the big car, such as a retired policeman and a security guard who as a boy borrowed his mother's Morris Minor, had a mild incident with it and wanting to return it as quickly as possible, rang Hunt's Garage. He remembers how a big recovery vehicle that sounded like an aeroplane came out to tow the little car home. Apparently at the time the police were helpful to local garages in dropping hints as to where crashed or broken-down cars needed succour and a fast breakdown truck could be useful in forestalling a rival garage's intentions!

Capt. Hunt had served in the RFC with 21 Squadron and had many friends in the flying world, including Capt. Broome who flew a Vickers Vimy to the Cape in 1920. He later joined Vickers Ltd and no doubt had heard of the Sunbeam-powered Mercedes that Hawker had built, and knowing it was no longer wanted would have acquired it and converted it into an exciting and powerful garage truck. His garage was taken over by Freyer's, and later became a Henley's garage. The Sunbeam-Mercedes is remembered by some of the people at Henley's as being replaced by a very early 40/50 hp Rolls-Royce, with crane mounted on its chassis, and apparently a Sunbeam was also used as a breakdown car by Hunt's Garage. But all trace of the car the great airman built has vanished, and its fate remains a mystery.

SIR E.T. SCARISBRICK'S 'SCARISCROW'

This aero-engined car probably appeared only once at Brooklands but is worthy of inclusion as having been mistaken for many years for a

'Chitty-Bang-Bang', even by its first American owner! It also had the distinction, according to Sir E.T. Scarisbrick Bt., of having a Benz engine from a German aeroplane shot down behind enemy lines during the war and taken out and buried by a friend for use later on, in a boat or a car. Presumably this drastic recovery of such a prized power-unit would not have been undertaken if it had been realized how easily and inexpensively such aeroplane engines could be obtained in England after the Armistice.

Thus it came about that the monster, later owned by Peter Helck, was bought unseen by Alistair Bradley-Martin of Long Island around 1938 because he was convinced, as were many other people, that it was a real, true, dyed-in-the-wool 'Chitty', created by the legendary Count Zborowski at his estate, Higham, at the Kentish village of Bridge, near Canterbury. He didn't exactly import a lemon—but he couldn't have been more wrong about the great car's pedigree.

The real 'Chittys' had been largely forgotten by then, such is the fickleness of memory. So when a car looking like a Zborowski 'Chitty' began to appear at speed-trials in the hands of G.P. Shea-Simonds, it was inevitable that this should be regarded as something out of Higham's past. Its owner himself was convinced that he had one of the ancient 'Chittys'.

The thing came to a head, as it were, soon after the outbreak of war. The late Laurence Pomeroy, who was then Technical Editor of *The Motor*, commenced therein a series of articles that recalled famous Brooklands' racing cars. In due course he described the 'Chittys', and stirred up a hornet's nest as to which was which and what had befallen them. Even Homer nods, and I see that Pomeroy quoted 'Chitty III' as a chain-drive car, which it wasn't, and was of the opinion that 'Chitty II' had gone to America, which at that time it hadn't. Having been told that there was a 'Chitty' in the USA, I too, fell for it being the second of Zborowski's cars. I knew that 'Chitty I' had long since been broken-up, after I had had a hand in the Conan-Doyle brothers bringing it to Brooklands in 1934 for a small exhibition of aged racing cars, where it had languished in the open, its scroll clutch useless, until John Morris bought it and cannibalized it, as he needed a gearbox for his 200 hp Benz.

It was only by digging about in the archives that I discovered what the aero-engined motor car in America at this time (1942) really was. It was hardly surprising that it was causing confusion.

I discovered that there had been another car, very similar indeed to 'Chitty II'. It had been built in 1921/22 to the order of the late Sir E.T. Scarisbrick Bt., of Banks, and in 1955 I was able to get in touch with him and learn more about it. (Incidentally, he owned many interesting cars, including the ex-John Cobb 2.3 litre Monza Alfa Romeo. The story of his aero-engined car was, however, romantic enough on its own.)

It seems that a friend of Scarisbrick's who was serving in France during the First World War came upon a German aeroplane that had landed behind the lines unharmed, due to running out of petrol. He thought the engine might serve his friend after the war for powering a motor boat. So he had it sawn out of the fuselage, carefully greased, and put into a packing-case, which was duly buried. The enemy recaptured the village where this was done at a later date, but never found the engine, which was afterwards disinterred in good order. (Incidentally, I think this may be the origin of the legend that the engine in 'Chitty I' had come from a shot-down zeppelin, whereas the late Col. Clive Gallop told me it was not of the type used in 'lighter-than-air' craft anyway, and I suspect it came from war-surplus stock, like so many others.)

When racing commenced again at Brooklands, Scarisbrick went to watch and was impressed when he saw the Count performing in 'Chitty I'. He decided that the engine his friend had so fortuitously obtained for him would be more fun in a car than in a boat. So he got C.H. Crowe & Co., of Kennington Lane, London, to install the engine in an old chain-drive 75 hp Mercedes chassis. This was a company which specialized in Mercedes repairs and such conversions—another was Aero Motors Ltd of Vauxhall Bridge Road, who had something to do with Ernest Eldridge's Isotta-Maybach, although I have never been able to discover any more about either of them.

The engine that Scarisbrick had acquired was a six-cylinder push-rod o.h.v. 145×190 mm. (18,882 cc) Benz. To set it in the suitably strengthened Mercedes frame the sump was removed and an oil tank on the outside of the chassis substituted, as done by Zborowski on his 'Chittys'. A tank carrying water for the transmission-brake matched the oil tank, on the opposite side of the chassis. The wheelbase was 11 ft. 6 in., and sprockets supplied by Mercedes-fancier Edward Mayer, giving a top gear ratio of 1.33 to 1 were used. Rudge Whitworth wire wheels and Hartford shock-absorbers figured in the revised specification. The original Benz twin carburettors fed through two manifolds of odd three-branch formation, on the offside of the engine, and on the nearside a new six-branch exhaust manifold fed into a huge tapering exhaust pipe. The noise apparently upset the Police when test runs were made along the Embankment. Two bucket seats were the only concession to bodywork, behind which was a stowage box. The original rear fuel tank was supplemented by a small scuttle-tank. Apart from being a two-seater whereas 'Chitty II' had four-seater bodywork, the Scarisbrick car, called the 'Scariscrow', was very similar in general appearance to 'Chitty II'; both originally had unpainted bonnets.

The car is said to have been fitted with a Bosch starter and to have cost about £1,600. The body was made by A.H. Story & Sons of Holloway in North London. Although the car became known as the 'Scariscrow' it was originally called 'Rabbit I', 'Rabbit' being the nickname of Sir Edward Scarisbrick's wife. The car was photographed in 1922 outside the National Gallery in Trafalgar Square, London, and in this picture the name can be seen inscribed on it. Perhaps this photograph was taken before one of the test runs along the Embankment to which the Metropolitan Police objected.

After seeing the 23-litre Maybach-engined 'Chitty I' performing at Brooklands, Scarisbrick probably realized that his smaller-engined car would be unable to compete in terms of lap-speed ('Chitty I' lapped at 113.45 m.p.h.) and although Zborowski was building 'Chitty II' with precisely the same Benz engine as Scarisbrick intended to use, and it was completed first, Scarisbrick was unaware of this. He was then living at Greaves Hall up in Lancashire, far from Zborowski, so would not have been conversant with the happenings at Higham. But the 1921 Brooklands season would have convinced him that his car would be no faster than 'Chitty II' (lap speed 108.27 m.p.h.) and thus would be rather overshadowed on the Track. The 'Scariscrow' was completed early in 1922 and, in lieu of racing at Brooklands, its owner decided to try it out at the Fanoe Island speed-trials in Denmark, remembering that John Duff and the 10-litre Fiat had made fastest time there the previous year.

Towards the end of the course at this beach venue, which was another preserve of aero-engined cars, including an Argus-powered Stoewer, one of the carburettors caught fire. But the Benz-Mercedes made fastest amateur time, and apparently won three other cups, clocking 101 m.p.h. over the kilometre. Scarisbrick's wife rode as passenger, and had a rather rough time, owing to the state of the beach. As a road car the Benz-Mercedes was satisfactory, except for bending the back-axle radius rods if it was accelerated too suddenly. But it was not used much (the tax was £79 per annum) and it was sold eventually 'to a Chinaman', it is rumoured for £200, whom Scarisbrick thought intended to put a closed body on it. But he never heard from this man again.

What happened to it in the interim, before it appeared in sprint events just prior to the war, is open to conjecture. When he discovered that the car was not, in fact, a 'Chitty', the American owner who restored the car in the 1950s issued a leaflet which attempted to clarify matters.

I think Robert Arbuthnot got hold of the ageing Scarisbrick car, when he was at Eton, and had a proper 2-seater pointed-tail body made for it, and that Shea-Simonds bought it from him. This gentleman, who was also an enthusiastic operator of a Speed-Six Bentley, ran the car in a speed-trial or two, and I think it was at such an event, probably at the dismal venue of the A.E.C. Works at Southall on the outskirts of London, that it ran off the course and damaged its old Mercedes radiator. To rectify this a Big-Six Bentley radiator was substituted, the bonnet being

Sir E.T. Scarisbrick's Benz-engined Mercedes in the Paddock at Brooklands. It had been inspired by 'Chitty-Bang-Bang II' and in later times was often mistaken for that legendary motor car. It never raced at the Track; the competition number probably belonged to a speed trial.

bodged to fit and the car daubed in red paint.

That was how the car looked when Bradley-Martin of Long Island imported it, in the belief that he had secured a 'Chitty'. Indeed, at the time he thought he had 'Chitty I', and when finally he was convinced that this 'Chitty' had long since been broken up in England, the excuse was made that a Benz engine looks very like a Maybach! For some time thereafter it was assumed that the car in America must be 'Chitty II'. This impression was enhanced by a photograph of the Scarisbrick car in the Paddock at Brooklands, although it was, in fact, wearing its Fanoe racing numbers. I believed this myself, until I was able to check that the latter 'Chitty' which I knew had been purchased by Mr Hollis of Dover from David Scott-Moncrieff, was still owned by Mr Hollis. It was then that I learned about the Scarisbrick car of 1922 and realized that that was what Shea-Simonds, and then Bradley Martin, had got hold of.

A 1919 Locomobile radiator was used to replace the Bentley radiator, George Rand and John Olivean doing much restoration, and the car was run in the V.M.C.C. of America's Easter Parade through New York in 1940, driven by Bob Heller, still posing as a 'Chitty'. A back-fire blew a hole in an intake manifold and the giant was ignominiously towed away behind a TA M.G. Walter C. Hadley of Connecticut then started some research to establish the car's true identity via Ken Taylor of T & T.'s, who enlisted the aid of Clive Gallop and myself in the process. The old car was sold in 1946 to the Ellis brothers of Sea Cliff, New York, and following a spell on blocks lasting for 15 years, was again restored. In December 1962 A.J. Hoe was instructed to rebuild the old car and Gus Reuter to repair the body. It had twin Zenith carburettors which Rand had fitted. The Benz-Mercedes then changed hands once more in 1949, and was now in the proud possession of famous US artist Peter Helck. It had rebuilt wire wheels to take 6.00×23 well-base tyres as fitted by Rand, and has a pointed-star badge labelled 'Benz-Mercedes', which was designed by Peter Helck. The cream body with red upholstery has been retained.

It is nice to know that a motor car in the 'Chitty' tradition is now properly preserved. But this story has a rather comic ending. As recounted earlier in this book, after the war Peter Harris-Mayes, to whom the real 'Chitty II' had been lent by Mr Hollis, and who had done so much good work on it, decided to accept an offer of £15,000 for it, from an American collector. All that Mr Hollis, backed by Lord Montagu, could legally do to retain the car in this country proved of no avail and it duly crossed the Atlantic. So, had Mr Bradley-Martin been able to wait, he might have acquired the motor car he thought he had acquired! As it is, America now possesses not only 'Chitty II' but the sister Benz-engined car that was for so long mistaken for a genuine Zborowski-inspired product.

There follow some American memories of the 'Scariscrow' by Smith Hempstone Oliver, as published in the *Veteran & Vintage* magazine Volume 21, Number 5, dated January 1977:

> I, too, while back in New York City in 1939 and 1940, thought that the Benz-engined racing car, then newly arrived there, was one of the legendary 'Chitty-Bang-Bangs', stories of which I had read many years earlier.

So it was with great delight that I amateurishly made an indoor snapshot of the great 18.8-litre engine in 1939 while the car was undergoing its restoration in the Rand establishment in the Brewster building in Long Island City. Here, while the intake manifolding had not yet been reinstalled, the newly painted cylinders of the rebuilt engine could be seen mounted on the scrubbed aluminium crankcase. I well recall how intriguing were the long polished push rods and the equally polished rocker-arm assemblies.

My next contact with the car came on the day of the very first Easter parade held in New York City by the recently formed Veteran Motor Car Club of America, this being on Sunday, March 24, 1940. A very cold day it was, and I now recall that while I was driving my 1914 Stutz Bearcat over the Brooklyn Bridge to the site of the start of the parade I was aware that the radio that morning had reported a temperature of a mere 15 degrees F., considerably below freezing. Five old cars participated that day, these consisting of the 'Scariscrow', driven by the late Bob Heller, three old pre-World War I shaft-driven Mercedes, driven respectively by Alec Ulmann, Charlie Stich, and George Rand, and my Stutz. Happy memories!

The remaining three photographs of mine were of Heller at the wheel of the 'Scariscrow' just before the start of the parade (with Tom Dewart in his non-parading M.G. alongside), Heller and Dewart in the 'Scariscrow' during the parade at 63rd Street and Park Avenue, and the stalled 'Scariscrow' later on at 52nd Sreet and 2nd Avenue. It was from this latter spot that the immense car was towed away by the tiny M.G. seen in the earlier photo. Steam can be seen emanating from the car, and I remember that Heller said the idling in traffic proved to be too much for the cooling system of the car, despite the temperature of the day. Also, the clutch had taken to slipping. After all, the intention away back in 1921 was not to create a passive vehicle suitable for New York City's traffic in 1940!

The enamelled Benz-Mercedes badge with the inverted 3-pointed star now seen on the Locomobile radiator, did not originate in Sir E.T. Scarisbrick's day, nor even by the time of the 1940 parade (as the photos testify), but is a product of that great artist, Peter Helck, the current owner of the 'Scariscrow'. At the time of the latest restoration of the car about a decade ago, this unique handsome badge was custom created to Mr Helck's design and attached to the radiator to identify the car to some and to confound the issue for others!

LT. COL. G.H. HENDERSON'S ROLLS-NAPIER

The aero-engined hybrid built for his personal transport by Lt. Col. G.H. Henderson would have been seen frequently at Brooklands, if as I think, he was the Col. Henderson who ran the well-known Henderson flying school there with 80 hp Avro-Renaults until it was taken over by Capt. Duncan Davis in 1928.

He put a 250 hp Rolls-Royce Falcon engine of just over 14-litres in a much modified 1907 40 hp Napier chassis. The Reg. No. was LX 9377. The engine came from a Bristol fighter, and others were obtained for use as spares. To obtain a straight-line transmission, engine and gearbox were inclined. The clutch and flywheel from the Napier were attached to the crankshaft of the Rolls engine, which was given Zenith carburettors and was started with a gigantic starting-handle, supplemented by a starting magneto. The rear fuel tank had once seen service in a Bristol fighter aeroplane and there was also a spare fuel tank.

The top-gear ratio was 2.2 to 1, which represented a maximum speed of 107 m.p.h. at 2,200 r.p.m., and 120 m.p.h. at 2,500 r.p.m., using 896 mm. tyres. If larger 935 mm. tyres were fitted a top pace of about 125 m.p.h. without exceeding 2,500 r.p.m. should have been possible. On the road, however, a speed of 74 m.p.h. was thought sufficient. At such speeds the petrol consumption was ten m.p.g.; one run of 375 miles at an average speed of 35 m.p.h. used up 33 gallons. Three and a half months after constructing his Rolls-Napier a mileage of 4,000 had been covered, in which no trouble had been experienced. The only servicing had been experimenting with different sparking-plugs and twice adjusting the valve clearances, which was described as an easy operation. The oil was changed once, in deference to loss from leaks which the owner had not troubled to stop. In those days frequent decarbonizing was a boring chore, but the Rolls-Royce engine showed no sign of over-carbonization and seemed likely not to require attention for 10,000 miles.

As for performance, the car was not regarded by its builder as a freak. What he had sought was the fastest general-purpose touring car on the road for the cheapest price. It would climb hills such as Wrotham, Reigate, Dashwood and Henley fast or slowly in top gear, and when an errant nut had damaged the gearbox Henderson drove for two weeks about Paris in the highest gear. The engine was reported to be easy to start 'single-handed' and to warm-up quickly. The cost of the Rolls-Napier in 1921 was a mere £650, which included engine, chassis, body, tanks, electric lighting, wheels, tyres, machining of parts, labour, painting etc. and some costly failures in the early stages.

Incidentally, in 1921 S.C.H. Davis of *The Autocar* had travelled in the car from Le Havre to Le Mans for the Grand Prix in 1921, done a lap or two of the course, then returned to Paris in it.

Henderson was anxious to meet someone who would be interested in co-operating in the building of an improved car using a 60 hp Napier chassis and he also thought one of his spare Falcon engines would make a good basis for 'A really fine' racing car. Perhaps, if this was the flying-school Henderson, aviation was not proving prosperous, because in 1924 he sold his car, after a single insertion in *The Autocar*, to H.W. White living just outside Glasgow, who was running a 1910 Rolls-Royce coupé. However, his enthusiasm for it was undiminished, and he described the delivery journey.

A stipulation of the sale was delivery of the car to Scotland. Henderson took the 250 hp Rolls-Napier out of retirement at Brooklands and went to Croydon in it to see the arrival there of the American fliers who had arrived from California in three Army Air Service Douglas bombers on an attempt to circumnavigate the world in 80 days, with 10,000 of the 20,000 miles still to do. He then left his home at Shepperton with his wife at 5.55 a.m. one Thursday morning in the summer of 1924 and got to Glasgow by 6.20 p.m., the 420 mile journey consuming 41 gallons of petrol. Stops of 1 hr. 5 min. gave a running time of 11 hr. 20 min. and, remember, the roads were nothing like as fast as those of the 1990s. The driver was slowed by heavy rain for the last 100 miles, and speed was held down because the tyres had not been changed for two years. No oil or water was added or needed on arrival. By this time the car's total mileage was 13,500 and it had given no trouble of any sort. The run to Scotland was done entirely in top gear, except for traffic stops, and the speed was held to 60 m.p.h., or 1,300 r.p.m. The car had a forward-mounted Rolls-Royce radiator, a six-branch exhaust manifold on each side, artillery wheels, and two aero-screens.

Around 1934/35 the car was used by R.G. Miller, who found it derelict in Skelmordie, where it had lain for some ten years. He fitted twin rear wheels from a Panhard-Levassor van and entered for the Bo'ness speed hill-climb in Scotland. On the way there the front universal-joint split at the keyway but was then welded-up. On its first ascent the Rolls-Napier made second-fastest time, but dirt in the petrol tank stopped it half way up the hill on its second run. The same trouble occurred at Stirling on the journey home. The old car was used occasionally on the road and was in the Glasgow area up to 1935.

THE 300 HP TIPO S76 FIAT

One of the biggest of the aero-engined cars, a quite outrageous monster, its engine capacity 28.4-litres, that made a fleeting appearance at Brooklands, although intended for attacking the Land Speed Record, was the 1910 Tipo S76 Fiat.

This remarkable car was built in 1910/11 presumably in the hope of taking the World Land Speed Record at around 140 m.p.h., this record then standing to the credit of Victor Hemery's Benz at 125.94 m.p.h. for a flying kilometre. This particular Fiat was certainly a clumsy, top-heavy vehicle. So high was the radiator that legend has it that you had to stand on the dumb-irons in order to fill it. In sober fact I believe the cap was over 5ft from the ground, and I have seen photographs which prove that fully-grown men, standing behind the car, could only just see across the bonnet.

The engine was one of the Tipo S76DA 300 hp airship power units, of 190×250 mm. (28,362 cc). It has been argued convincingly by Pomeroy and others that the Tipo S61 Fiat GP engine of 1909 was an Ettore Bugatti design and, if so, presumably this monobloc overhead-camshaft 16-valve dirigible engine had the same origin. The very long stroke of this engine and its o.h.c. valve gear necessitated a very tall bonnet, the radiator shell curving upwards to meet it, and the tall filler cap protruding through the shell. Yet, in spite of the imposing height and size of the engine, which apparently was installed in the car unmodified from constant-speed form, developing maximum power at 2,200 r.p.m., the chassis was flimsy in the extreme and could well have been a 1907/08 frame, especially as the wheelbase was unusually short in relation to the length of the bonnet. The car had full-length undershields, typical Fiat outside brake and gear levers and drilled sprockets, while the tail was not unlike the one Duff used on the resuscitated 'Mephistopheles'.

The drive went via a Hele-Shaw multi-plate clutch to a massive gearbox, and final drive was by side chains. Wire wheels were used and an airship-tailed two-seater body completed the wicked ensemble. The occupants sat in extremely high seats and the exhaust issued from two rectangular stubs on the nearside. The front dumbirons and axle carried

The 28.4-litre 190×250 mm. (28,362 cc) overhead camshaft Tipo S76DA Fiat engine intended for the Forlanini airship, said to develop 290 b.h.p. at 1,900 r.p.m., and used for the astonishing 300 hp Fiat racing car.

streamlined fairings.

Having built the monster, Fiat seem to have been puzzled as to its destiny. At all events, Mr D'Arcy Baker arranged for it to appear at Brooklands in Felice Nazzaro's hands, and appear at the Track it did, although Petro Bordino deputized for Nazzaro. This was at the conclusion of the 1911 Whitsun Meeting, when the giant Fiat did a couple of demonstration laps but, hampered by barriers placed on the course, did not exceed 90 m.p.h. In any case the Fiat's appearance was not billed in the programme and it is probable that the car was being prepared for the Saltburn sand races and was not giving a serious demonstration. (The official Fiat history states that 'Bordino beat the mile record at Brooklands in a 300 hp car' in 1911 but there is no record of this in the Brooklands archives.)

A reliable eye-witness has told of how Bordino, like some other Continental drivers new to Brooklands Track, disliked using the bankings and how the sight of this top-heavy car lapping low down, with flames

The biggest of them all! The ungainly and excessively tall 300 hp Tipo S76 Fiat that came to Brooklands in 1911, in the hands of Bordino, and also ran at Saltburn. It was really intended for an attack on the Land Speed Record, which it never achieved.

belching from its open exhausts and a roar of sound, added to the excitement. No wonder that in certain quarters this enormous Fiat was referred to as 'The Beast of Turin'.

The Fiat duly arrived in Yorkshire. It seems that in practice at Saltburn, after sinking into soft sand and needing many horses to retrieve it, Bordino was timed unofficially at 125 m.p.h. On race day, however, with the course in poor shape, he managed only 116 m.p.h., although this, it seems, was a very thrilling spectacle. The official Fiat history exaggerates this to a record of 154 m.p.h.!

After this brief appearance in England the 300 hp Fiat returned to Turin. It might have languished there for ever had not a Russian Prince taken a fancy to it. He bought the car and engaged Arthur Duray to drive it on the long-delayed assault on the Land Speed Record. Duray, perhaps not surprisingly, for he was a Belgian, decided to use a 5-mile straight road near Ostend. Over this course, in December 1913, under adverse weather conditions, he was timed officially at 132.37 m.p.h. over the flying kilometre. Since 1910 the A.I.A.C.R. had required two runs, in opposite directions, to be made and for some reason unknown to me Duray never established a time for a second run.

However, the Russian decided to return in the Spring for another attempt and to this end various vital spares were stored in Ostend, although the Fiat returned to Turin. The war over, Duray is said to have returned to Ostend, only to find all the spares missing, presumably either confiscated or commandeered by the German Army during the occupation of the town. The Russian sponsor also failed to survive the war, vanishing without trace, perhaps assassinated in the revolution of 1917. That is the story W.F. Bradley of *The Autocar* told in 1925, but if

Duray went in search of the spares it seems that either he or Fiat had hoped to run the car on their own account and it is a mystery why fresh parts could not have been found, or made, by the factory.

The fact remains that the giant Fiat was still at Turin after the first Tipo 501 was built, for it figured in a publicity picture standing beside one of these touring cars in 1920. It is said finally to have been bought by someone in Mexico and to have been sent out to Tampico at this person's request, never to be heard of again.

AND SOME OTHERS

Another car powered by an aeroplane engine must have come to Brooklands, because its owner, described how it had been timed over the half-mile at the Track at 106.5 m.p.h. and how while there he started from rest at the foot of the 1 in 4/1 in 5 Test Hill in second gear, changed into top gear after five yards, and the car accelerated all the way to the summit. The engine was a 200 hp vee-eight Hispano Suiza and I think it may have been in an RFC Crossley tender chassis, built at the RAE Farnborough during the war.

The car was geared for 100 m.p.h., gave a regular 15 m.p.g. of petrol, or 18 m.p.g. when used for touring in France, Italy and Switzerland. It was equally at home in top gear in London traffic, and the acceleration to high speeds was described as 'really astounding', making for excellent average speeds. The engine was taken down, out of curiosity, after 10,000 miles and nothing was worn, but over-oiling had resulted in some carboning-up. To start the engine a starting-handle geared at 10 to 1 was used in conjunction with a starting magneto geared to the handle and this enabled a boy of seven to crank it up, and the two B.T.H. polar-inductor magnetos, set close up behind the radiator, so inaccessible, had proved trouble-free. This may have been the car owned by Capt. J.V. Nash in 1921, who took the engine from an SE5a fighter and retained the

The 21-litre Metallurgique-Maybach as restored by Douglas Fitzpatrick, with its rakish three-seater body.

Crossley gearbox and, with a top-gear ratio of 2.25 to 1, got a petrol consumption of 15 m.p.g.

Another aero-engined car which would presumably have been seen at Brooklands was that owned at one time by the late Lord Donegall, who when writing his (unpublished) autobiography, remembered that he bought it at Brooklands, but not much else about it. According to the memory of the late Leo Villa, this was an Itala chassis into which Giulio Foresti, who looked after Itala interests in this country, tried to install a 250 hp six-cylinder Hall-Scott aero-engine, around the year 1916. He had difficulty in mating the flywheel and clutch to the crankshaft of the American aero-engine and when the propeller-shaft came away at the front and dug into the road, carrying away the brake gear, the project was abandoned. It is said that the car was eventually towed down to Count Zborowski's place in Kent, which may be why Lord Donegall thought his car was one of the Count's racing cars.

Another aero-engined car which may have disported itself on the Weybridge concrete is the impressive 21-litre Metallurgique-Maybach owned by the late Douglas Fitzpatrick. I have no proof that it did, but it was rumoured that after Mr Cole of Brundall, near Norwich and his partner Mr Tillet had put the Maybach engine into the 1906/07 60/80 hp Metallurgique chassis in 1919 it did 118 m.p.h. on Brooklands Track. There is no reason why it should not have done this, although Weybridge is a long way from Norfolk; and other accounts say that Mr Cole died in 1920, before using the hybrid.

It is remarkable that after Mr Cole's death the hybrid lay undiscovered for many years in a barn close to Fitzpatrick's house, Sheringham Hall. He heard of it, fortunately, took it over, and restored it to his usual high standards, aided by a German prisoner-of-war, whom he was able to retain to look after the needs of the Maybach engine. This was a 180 hp T-head 160×170 mm. Series-AZ, No. 30, the first type of engine of its kind built by Maybach, prior to 1912. This model was superseded by the Series-CX, rated at 210 hp, which had a longer piston stroke and a more rigid crankcase with six-point instead of four-point mounting, and the timing gears transferred to the back of the engine. On first encountering the moribund giant car in its shed, Douglas Fitzpatrick said: 'I wasn't prepared for the sight when I opened the door. Even Aladdin, when the door of the Mountain Cave rolled open to display the jars of shining gold within, could not have gone through a greater emotional crisis than I, when I looked in the shed and saw this monster car, of a species I couldn't even recognize.'

It appears that E.A.D. Eldridge owned the Metallurgique before the Maybach engine was put into it, which apparently necessitated lengthening the chassis by 18 inches. Fitzpatrick had a three-seater body made for it and, as I know from personal experience, the performance is very impressive. The new owner took the car to various vintage meetings, including a Brooklands Society Reunion. After his death it was to be auctioned, but broke the bolts holding the flywheel to the crankshaft. However, repairs were effected in time for the new owner, V.C.C. and V.S.C.C. member Brian Moore, to enter it for the V.S.C.C.'s 1990 Weston-Super-Mare speed-trials, to which it was successfully driven from Cambridge, but was a non-starter due to water getting into No. 3 cylinder. In 1922 it won the Edwardian class (20.74 sec. for the s.s. half-kilometre).

Finally, we come to the Land Speed Record cars. These have used a great many different aero-engines, but they have been well documented elsewhere, particularly in *Land Speed Record* by Cyril Posthumus and David Tremayne (Osprey, 1985) and in my own book on the subject, published by Motor Racing Publications in 1951 and revised in 1964; and unless they ran at Brooklands or were built there they are outside the scope of this book.

Apart from those which took part in Brooklands races, as already covered, the first car to officially exceed 200 m.p.h., the '1000 hp' twin-engined 44-litre Sunbeam, driven at Daytona, USA, in 1927 by Major (later Sir) H.O.D. Segrave, was demonstrated before the Brooklands' spectators at the B.A.R.C. Whit-Monday meeting of 1927, having been on display on the tennis court adjacent to the Paddock during the afternoon. It was timed to happen at 4.10 p.m., and in rain, which caused the rest of the programme to be cancelled, the bareheaded Segrave went ahead, using, it is believed, only one engine. Even so, the tail of the monster red car came round whenever he opened-up. In a series of

Preparing Sir Malcolm Campbell's L.S.R. Napier-Campbell 'Bluebird' for its Brooklands demonstration.

swerves Segrave drove two clockwise laps round the public enclosure, to applause from those huddled under their umbrellas.

In 1928 it was the turn of Capt. Malcolm Campbell (also not yet knighted) who maintained premises in the Brooklands Paddock, to display his 450 hp Napier-Campbell 'Bluebird' which had set the L.S.R. to 206.95 m.p.h. at Daytona that February, at the Easter Monday meeting, at which he did three quite fast laps of the Mountain circuit, before the competitors in the last race of the afternoon lined up. At the following charity meeting 'Bluebird' was shown for a shilling (5p) a time, in the Paddock—proceeds going to Westminster Children's Hospital, so some good came from all this aero-engined motoring.

Another demonstration, of Malcolm Campbell's Rolls-Royce 'Bluebird', was given at the 1933 Brooklands Easter Monday meeting, when his Rolls-Royce-Campbell, which had lifted the L.S.R. to 272.46 m.p.h. actually did an erratic lap of the outer circuit, after being pushed out and started up with an auxiliary engine amid flames and black smoke. So the spectators had the unique opportunity of seeing the fastest car in the world tackling the bankings, quite a difficult procedure. The last of these demonstrations was made at the 1935 Easter Monday meeting, when Campbell showed off the twin-rear-wheeled 'Bluebird' in 276.82 m.p.h. form, powered with that 36.5-litre 2,300 hp supercharged R-type Rolls-Royce racing seaplane engine.

Campbell had engaged Thomson & Taylor, the Brooklands engineers,

Continued on page 153

The Napier-Campbell about to leave the Paddock.

In the Finishing Straight, watched by an expectant crowd. The Ace wheel-discs and the lifting jacks can be seen.

Campbell leaving on his demonstration run at the Track, at Easter 1932. It is said that although only out for a brief run the concrete stripped the wafer-thin special tyre treads.

HOMSON & TAYLO

Above *An unusual sight! Campbell's L.S.R. 'Bluebird' on the Members' banking.*

Left *In 1935 Sir Malcolm Campbell again treated the Brooklands' spectators to the spectacle of 'Bluebird' running on the banking, this time with the Rolls-Royce-Campbell on its return from Daytona, where it had lifted the Land Speed Record to 276.82 m.p.h.*

Opposite *Two views of 'Bluebird' running up the Brooklands' Finishing Straight.*

Right *When John Cobb decided to go for the L.S.R., Reid Railton designed for him the ingenious Railton Special, powered with two Napier 'Lion' aero-engines. It was built by Thomson & Taylor's at Brooklands, and was demonstrated to the Press outside the workshops, which have since been destroyed, and is seen being manhandled out of the sheds. A Terraplane stands in the background, the car from which Railton developed the sports Railtons.*

Below left *With the newsreel camera at the ready, those who helped to build the Railton duly admire it; Kenneth Thomson of T & T's is standing arms akimbo.*

Left *How one of the Napier 'Lion' engines was installed in the ingenious Z-shape chassis.*

Below *The body of the Railton Mobil Special about to be lowered on to the chassis, as Cobb sits and waits. The old Brooklands sheds on the Byfleet side of the estate form the background.*

The Railton Mobil with its sleek body in place outside the old Parry Thomas premises at Brooklands.

John Cobb in the Railton Mobil Special, with which he set the Land Speed Record to 369.70 m.p.h. in 1939 and to 394.20 m.p.h. in 1947.

A test rig used at Brooklands to experiment with aeroplane engines and propellers. The chassis is thought to have been built up from Minerva parts.

to modify and re-engine 'Bluebird' for the later L.S.R. attempts, so apart from these public demonstrations on the Track the car would have been shown to the Press outside their premises on the Byfleet side. John Cobb's remarkable twin-engined Napier Mobil Special which raised the Land Speed Record at Bonneville, USA to 357.20 m.p.h. in 1938, to 369.70 m.p.h. in 1939 and to 394.20 m.p.h. in 1947, was also built at T & T's Brooklands workshops, with Reid A. Railton again responsible for the design work. So this car, too, would have been seen outside the old Parry Thomas workshops, where it would be photographed by the Press and its monocoque body shell tried out on the unique Z-shaped backbone chassis, and shown to the newspaper men, although this L.S.R. car was not run on the Track.

Alas, in 1978 the T & T workshops, Parry Thomas's bungalow 'The Hermitage', and all the other sheds which formed the line of the buildings facing the Aerodrome—the entire 'flying village'—were bulldozed to rubble with little prior warning, to make way for Bass Charrington's, the brewers, new premises. Since that happened, and other famous landmarks of Brooklands have been demolished, the old Motor Course has lost much of its attraction for me. Would that we could return to the days when the aero-engined monsters battled it out in the traditional handicap contests, at a place which should have become a carefully-preserved and cherished British historical heirloom.

Addendum

ENGINE AND CHASSIS NUMBERS

When asked to try to identify historic cars I make a habit of asking their owners to quote the relevant numbers to me, rather than the other way about, for obvious reasons! However, the aero-engined cult being virtually dead, except for Peter Morley's well-known 24-litre Bentley-Napier in V.S.C.C. events, there seems to be no harm in quoting what numbers I have from the Brooklands archives for some of the racing cars of this type, just for the record:

	Engine No.	*Chassis No.*
1913 V12 Sunbeam	No numbers were declared, even for record-attempts.	
1919 350 hp Sunbeam	1	1
'Chitty-Bang-Bang I'	1738	10216
'Chitty-Bang-Bang II'	34769	not quoted
The Martin-Arab	79059 and 70957	not quoted
The Cooper-Clerget	205	30071
The Wolseley-Viper	not quoted	2482
The Isotta-Maybach	M73	not quoted
The Fiat 'Mephistopheles'	20270	not quoted
The 'White Mercedes'	MN/4227	25655
The Higham Special	S2	not quoted

Index